bizmanualz®

Disaster Recovery Procedures for Business Continuity Management

The Professional's Ready-to-Use Guide for Disaster Preparation and Response

Bizmanualz, Inc.

St. Louis, MO, USA

bizmanualz®

TRADEMARKS

Bizmanualz® is a Trademark of Bizmanualz, Inc.

This publication is sold with the understanding that the publisher is not engaged in rendering legal, accounting, or other professional services. If legal advice or other expert assistance is required, the services of a competent professional should be sought. The Publisher cannot in any way guarantee the forms, agreements and statements in this manual are being used for the purpose intended and therefore assume no responsibility for their proper and correct use.

Printed in the United States of America

ISBN-13 978-1-931591-40-9

For more information about available fully editable Policy and Procedure publications visit Bizmanualz on-line at www.bizmanualz.com.

Others in the Bizmanualz Professional's Ready-to-Use Procedure Series include:

Accounting & Bookkeeping Procedures for Internal Control
Computer & Network Procedures to Manage IT Systems
Finance & Treasury Procedures for Compliance and Performance
Sales & Marketing Procedures for Sales Pipeline Management
Business Security Procedures to Protect Assets and Employees
Human Resources Procedures for Employee Management
ISO 9001 Standard Procedures for a Quality Management System

Cover illustration by Leo Kundas

Disaster Recovery
Policies, Procedures and Forms

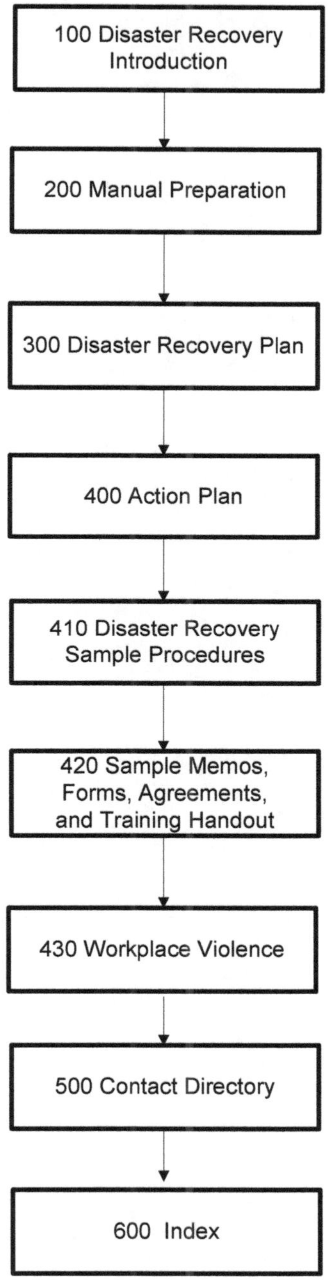

[This page intentionally left blank]

DISASTER RECOVERY

Introduction
100

[This page intentionally left blank]

DISASTER RECOVERY

Table of Contents

Tab Title ....page

[This page intentionally left blank]

1.0 Disaster Management and Business Recovery

A disaster is any event that will significantly impact the company's operations - Anything. Traditionally, this means fire, flood, and earthquake. Non-traditionally, this means a toxic waste dispersion, computer system crash, power outage, labor strike, community illness and many others.

No one is exempt from a disaster, and most people are unprepared if a disaster erupts. The company has prepared as if the disaster will exceed our ability to respond effectively: a worst-case scenario.

Disasters result from three (3) types of incidents, caused by:

1. Natural or cataclysmic events (i.e., earthquake, fire, flood, storm);
2. Human behavior (i.e., robbery, bomb threat, arson, hostage event, transportation strike); and
3. Technological breakdowns (i.e., power outage, computer crash).

The key to disaster management is to have a plan in place when the disaster strikes. A set of simple, effective guidelines and procedures for all people to follow. Just as a ship without a rudder is at the mercy of the tides, a company without a plan is at the mercy of events.

Human beings tend to make inappropriate decisions during a crisis. The company has a plan for coping with most emergencies already prepared and shared with employees, and we stand a better chance of surviving any emergency and recovering rapidly.

The emergency we have prepared for is one that will cause one or more of the following results to occur in our communities:

- The governor of the state or another competent authority declares the community a disaster area.
- Martial law is declared, the National Guard is activated, and basic civil rights are suspended.
- A state of total lawlessness in the community; law enforcement agencies are unable to cope with the results, even with the assistance of the National Guard.
- One third of the community dies immediately, or becomes so gravely disabled that they are unable to care for themselves or others.
- One third of the community is unavailable, unable or unwilling to help in disaster recovery efforts.
- One third of the community becomes responsible for the welfare of the remainder of the community, assuming roles and responsibilities for which they may be untrained and unprepared to handle.

Traditional roles and routine activities within the community are temporarily suspended. This causes the reassignment of personnel to fulfill necessary roles and functions, with temporary powers, duties and responsibilities.

A 72-hour interruption of:

Basic emergency and transportation services, such as police, fire and medical assistance, and both public and private forms of transportation. Routine calls for service are not answered, and the agencies' priorities will involve life-threatening situations only.

Food, water and survival supplies. Deliveries to food stores and hospitals are interrupted or hijacked.

People immediately strip existing supplies of all kinds, stockpiling their own reserves. Water pipes break, crack or clog, reducing or eliminating water flow.

Electrical power, sanitation services, sewer. Power lines are destroyed, overloaded or collapse. Water flow is restricted, and sewer pipes break, crack or clog and back up. Effluent reservoirs burst and flood the surrounding area with untreated waste.

Vendors of products and services will be unable to fulfill their roles, either because the geographical area affected by the disaster is closed to outside traffic; phone lines are down and information cannot be shared; or the vendors will be too busy answering calls for service from other clients also affected by the disaster.

We have planed for a disaster as if we will not have any outside help for at least three to five days. We should be self-sufficient during that minimum amount of time, and preferably longer.

2.0 Goals of Disaster Management and Business Recovery

The three primary goals of disaster management and business recovery planning are to:

1. Eliminate or reduce the potential for injuries or the loss of human life, damage to facilities, and loss of assets and records. This has required a comprehensive assessment of each department within the company, to insure that appropriate steps have been taken to:
    - Minimize disruptions of services to the company and its customers;
    - Minimize financial loss;
    - Provide for a timely resumption of operations in the event of a disaster; and
    - Reduce or limit exposure to potential liability claims filed against the company, and its directors, officers and other personnel.
2. Immediately invoke the emergency provisions of Disaster Management Plan to stabilize the effects of the disaster, allowing for appropriate assessment and the beginning of recovery efforts. We then contain the disaster to its least impact and provide for the fastest recovery.

3. Implement the procedures contained in the Disaster Management Plan according to the type and impact of the disaster. When we implement these procedures, we have prioritized all recovery efforts as follows:

    - Employees: Not only must we help to ensure their survival as a basic human concern, but for their performance in helping other persons on company premises when the disaster strikes.
    - Customers: As we do with employees, we must help to ensure the survival or care of customers affected by the disaster; physically, mentally, emotionally and financially.
    - Facilities: After ensuring the safety of employees and customers, we then secure the facility as shelter for both people and assets.
    - Assets: Although for the most part recoverable, assets will only be addressed after people and facilities are secure.
    - Records: As with assets, we will only address the recovery and reconstruction of important records when all people are cared for, facilities secured, and all assets have been audited and stored.

3.0 The Disaster Recovery Plan

Plan Effectiveness

In order to be effective, the company's disaster management and business recovery plan contains:

- The company's philosophy, mission statement and goals regarding disaster management and business recovery.
- Written and approved executive succession instructions.
- The appointment of a temporary Disaster Management Executive Committee for the term of the emergency, who may also act in the absence of the company's Board of Directors.
- Clearly defined guidelines and scope of all disaster management and business recovery efforts, based upon a thorough risk-assessment exercise.
- Clearly defined duties, authority and responsibilities for each employee classification, with designated primary and alternate department leaders and staff personnel to manage critical functions.
- A business recovery plan (operations manual) for each office, department, facility and function within the company, and for essential service vendors.
- Designated and equipped sites for assembly of personnel for each phase of the disaster management and business recovery effort.
- A well-documented testing and evaluation process to be conducted at specified intervals, and at least annually.
- A comprehensive training program for all personnel.

- Written copies of the final Disaster Management Plan distributed to office and department leaders, including a complete list of all emergency response agencies and facilities.

The Disaster Recovery Team

The company's Disaster Management Team is drawn from the company's available personnel resources, and may include service vendors. The Team functions as a strategic planning and tactical response unit. Team members may be assigned temporary duties, responsibilities and a level of authority beyond their normal employment classification.

The Disaster Management Team is comprised of executives, managers and department heads with functional responsibilities involving several offices, departments or facilities. The Team may include sub-groups as appropriate.
The Disaster Management Executive Committee is charged with the responsibility for creating a comprehensive Disaster Management Plan for the company. The Committee is comprised of the company's President/CEO, selected executives and senior managers, and may contain members of the company's Board of Directors. The Disaster Management Plan is used as the guiding mechanism for fulfilling the company's duties and responsibilities regarding local, state and federal regulations.
The Disaster Management Team Chairpersons are responsible for assigning the research, development and implementation of the Disaster Management Plan throughout the organization.
The Chairpersons are also responsible for assuming control over business operations during and after the disaster, until business operations return to normal. The Chairpersons are responsible for providing effective leadership and administration of the company's recovery efforts.
The Disaster Management Team Coordinators are charged with the responsibility for creating a comprehensive Disaster Management Plan for the company, and for implementing the decisions made by the Chairpersons and the Disaster Management Executive Committee during a disaster. The Coordinators are responsible for providing effective management and implementation of the company's recovery efforts.
The Disaster Management Team Department/Office Leaders and service vendors are responsible for acting as Coordinators for their respective areas of responsibility, and report directly to the Disaster Management Team Coordinators.

Synopsis of the Disaster Recovery Plan

The Disaster Management Plan in summary:

Sections 1.0 through 3.0 address the procedure for the formulation of the Company's policy and procedure mechanism relating only to the plan an overview to the concept of Disaster Planning, and the company's statement of purpose regarding emergency activation of any of the measures included in the plan.

Section 4.0 Disaster Management Team. Identify team members and their responsibilities.

Section 5.0 Notification of Emergency Responsibilities. Team members must be notified of their duties, attending meetings and remain up-to-date regarding company policies.

Sections 6.1 through 6.7 identify all personnel responsible for the initial and continuing research, development and implementation of the plan, establishes the formal notification and reporting structure for all members of the Disaster Management Team, describes the duties and responsibilities of all personnel and assigns appropriate levels of authority to those personnel.

Section 7.0 describes the alert scenario for differing levels of emergency, and the stages of emergency used to calculate the most effective response.

Section 8.0 identifies the Emergency Response Procedures that all employees must be able to identify, prioritize and act upon to diminish the impact of common events.

Section 9.0 addresses agreements for services and agencies responsible for coordinating disaster management and business recovery efforts in the community.

Section 10.0 identifies the site of primary and secondary Centers of Operations, to be used in the event this plan is activated, an itemized list of supplies and equipment to be distributed to these centers of Operations, and itemized list of supplies and equipment to be stored at each office, and the itemized list of supplies and equipment to be stored at the storage site, if appropriate.

Section 11 describes the testing and training requirements to be developed upon the Board's approval of the plan. This testing and training program has been developed to address the company's entire staff regarding policies, procedures, equipment and other topics to be developed.

Section 12.0 describes the annual evaluation process to be conducted by each department. The annual report, to be issued by the Disaster Management Team Coordinators and approved by the Board, must address the status of each department regarding compliance with established policies and procedures, a projection regarding any new equipment, personnel, policies or procedures required for the coming year, and suggested changes to existing policies or procedures with accompanying rationale.

Section 13.0 through 16.0 contains each department's and offices recovery plan for resumption of normal services after the disaster. This section includes identifying information on all personnel, equipment, facilities and supplies, disaster evacuation, assembly and notification procedures, and communications information.

Section 17.0 contains the training program, to be supplemented with vendor and related information as appropriate.

Section 18.0 is reserved for information added to the plan and documented changes at intervals.
Section 19.0 contains the forms to be used by the company during and emergency activation of the plan.

Section 20.0 is the actual resolution authorized by the Board, and insures the continuity of management and business operations by providing emergency operating policies and procedures, description of who may declare a disaster leading to the implementation of the plan, assignment of powers, reporting structure, responsibilities and authority for emergency action; provisions for removal of emergency conditions; and the description of locations and powers of temporary main offices.

Sections 21.0 and 22.0 identify the individual members of the Disaster Management Executive Committee; assigns powers, authority and responsibilities to individual members; and provides for the termination of status. Section 21 is the appointment document, and Section 22 is the acknowledgement made by each Committee member.

Section 23.0 is the resolution signed by the Board's Secretary upon the Board's approval, authorizing The Disaster Management Plan as the company's only plan, and authorizing the Disaster Management Team Chairpersons and Coordinators to carry out all provisions of the plan.

Section 24.0 is the annual re-certification resolution of the Disaster Management Plan based on the recommendation from the Disaster Management Team Coordinators and the Disaster Management Team Chairpersons.

Section 25.0 provides for the storage of resolutions and succession documentation, and access control codes and keys for facilities.

Section 26.0 contains listing of the confidential information retained in a separate binder, and stored with the Disaster Management Plan. This is not to be included in copies of the Disaster Management Plan distributed to individual offices and departments.

Section 27.0 Contains the LEADERS TRAINING GUIDE used to inform all personnel about the company's DISASTER MANAGEMENT PLAN.

4.0 Action Plan

This section provides the detail steps that are recommended to permit the development of a complete Disaster Management Plan. The Action Plan is presented in the section that follows the "ACTION PLAN" tab.

5.0 Examination and Audit Guidelines

The Examination and Audits section provides a checklist of basic questions for the entire organization. These basic issues need to be addressed in order to provide the ability to recover from a disaster.

6.0 Disaster Response Guidelines

If a disaster occurs, the discovering employee is to promptly notify his/her immediate supervisor. If the immediate supervisor determines that activation of the plan is required, he/she will promptly:

- Notify Coordinators and Chairpersons, if the disaster is discovered during business hours; or
- Notify persons at the selected emergency response phone number, if the disaster is discovered during non-business hours.

Persons at the selected emergency response phone number maintain emergency phone numbers for Chairpersons, Coordinators and members of the Disaster Management Executive Committee.

Persons at the selected emergency response phone number will initiate appropriate emergency notifications, and will also initiate an activity log to track events relating to its role in the recovery effort.

Upon notification by either the reporting supervisor or the Coordinators, the Chairpersons will formally initiate the Disaster Management Plan, and authorize appropriate emergency response to the designated Centers of Operations (Command Centers):

Chairpersons, Coordinators and appropriate Regional, Office and Department Managers will initiate independent activity logs, to track events and their respective roles during the recovery efforts.

The Disaster Management Executive Committee will notify available Board members, at a time and in a manner the Chairpersons consider appropriate.

The Disaster Management Executive Committee will assemble at the designated Command Centers as assigned.

Coordinators will gather data for the initial damage assessment. Based upon this initial assessment, the Disaster Management Executive Committee will determine the appropriate level and method of response.

Coordinators will assign emergency contact with phone system representatives to appropriate staff, who will notify company representatives of the disaster.

Emergency personnel will be assigned for telephone staffing. All personnel assigned to staffing emergency telephones will:

- Assemble at the designated sites as assigned;
- Initiate independent telephone activity logs to track events and their respective roles during recovery efforts;
- Maintain designated emergency:
- Lines for contact with Chairpersons and
- Coordinators; members of the Disaster Management Executive Committee; and appropriate Office and Department Managers;
- Lines for employees; and
- Lines for routine business and other business functions as available.

The Disaster Management Executive Committee will notify appropriate Regional, Department and Office Managers to respond promptly to designated Command Centers, or other appropriate sites.

Emergency personnel will be assigned for on-site telephone staffing. All appropriate Regional, Department and Office Managers and staff personnel assigned to staffing emergency telephones will:

- Assemble at the designated sites as assigned;
- Initiate independent telephone activity logs to track events and their respective roles during recovery efforts;
- Maintain designated emergency:
- Lines for contact with Chairpersons and
- Coordinators; members of the Disaster Management Executive Committee; and appropriate Office and Department Managers;
- Lines for employees; and
- Lines for routine business and other business functions as available.

Coordinators will manage responses by personnel, vendors, Regional, Department and Office Managers from the designated Command Centers. Regional, Department and Office Managers will manage on-site responses by the region's emergency services agencies, and other on-site functions as assigned by the Coordinators, including:

- If facilities or assets only have been damaged, representatives from the facilities management, insurance and security functions will be dispatched for on-site damage assessment and control.
- If records have also been damaged or compromised, a legal representative will be dispatched for on- site damage assessment and control.

- If the Data Processing facility has been damaged or is inaccessible, appropriate functions will be shifted to the designated recovery site(s).
- If records have also been damaged or compromised, a legal representative will be dispatched for damage assessment and control.

Based upon the on-site damage assessment, the Disaster Management Executive Committee will determine whether to:

- Close the facility, office or department, and
- temporarily re-assign employees and customer service functions;
- Open the facility, office or department on limited basis, and advise personnel and customers; or
- Open facility, office or department on full-service basis.

Based upon on-site damage assessment and the course of action determined by the Disaster Management Executive Committee, the Chairpersons will initiate an initial press conference and issue a press release. This press release will include statements regarding:

- An initial damage assessment to facilities only:
- Statements regarding injuries, fatalities, and general geographic conditions will be referred to the responsible emergency services agencies;
- The name(s) of the person(s) responsible for managing the recovery effort;
- The names of persons authorized to release information to the media;
- If credit functions are affected, that all customer accounts are safe and protected, and when the credits to those accounts will be available; and
- The location designated as a temporary customer-service center, including:
- The name of the on-site representative(s);
- The physical address of the location;
- Any "hotline" number designated for customer inquiries; and
- The hours of operation.

The Disaster Management Executive Committee will initiate the shift from recovery (stabilization) to restoration (of critical functions only), resumption (of all functions), or reconstruction (to the point of disaster). This includes:

- The notification of any appropriate regulatory agencies and insurance carriers of the disaster and a summary of actions taken, if appropriate; and
- The issuance of periodic press releases by the Chairpersons, as appropriate.

Based upon the continuing damage assessment, progress reports and recovery efforts, the Disaster Management Executive Committee will determine when to de-activate the disaster status. When the shift to routine operations is initiated:

- Interim policies and procedures will be implemented as appropriate;
- Continuing restoration efforts will be delegated to the on-site representative(s);
- Emergency telephone personnel will be assigned to contact appropriate Regional, Office and Department Managers, and persons at the selected emergency response phone number with instructions de-activation instructions;
- The Disaster Management Executive Committee will conduct its own de-briefing and critique immediately following the plan's de-activation, and:
- The Chairpersons will issue a press release summary of actions taken and disaster status;
- Notify any appropriate regulatory agencies and insurance carriers of the return to routine operations, or continued restricted activities, if appropriate; and
- Advise the Board of Directors of appropriate recovery efforts.
- Appropriate Regional, Office and Department Managers will furnish a written after-action report within 24 hours of the plan's de-activation; and
- All independent activity logs will be delivered to Command Center, as attachments to the after-action reports written by Regional, Office and Department Managers.
- Within 24 hours of the plan's de-activation, the Auditor will critique the plan's implementation and:
- Meet with the Disaster Management Executive Committee and discuss appropriate findings; and
- Make recommendations to revise the Disaster Management Plan, if appropriate.
- Within 24 hours of the Audit review, the Chairpersons will:
    - Notify any appropriate regulatory agencies and insurance carriers of continuing recovery efforts and a summary of actions taken, if appropriate; and
    - Prepare and distribute a written summary of the disaster and recovery efforts to any appropriate regulatory agencies, insurance carriers and appropriate personnel, including the Board of Directors.

DISASTER RECOVERY

Manual Preparation

200

It is vital in order for your manual to be effective that it be clearly written and easily understood by all the employees in your organization. This section provides an introduction to producing your manual

[This page intentionally left blank]

Manual Preparation
Table of Contents*

* The Manual Preparation Table of Contents references section page numbers (lower right corner).

[This page intentionally left blank]

SECTION 1 INTRODUCTION

This Disaster Recovery Manual was developed to assist organizations in preparing a Company Disaster Recovery Program. It can be modified to fit one's individual company concerns and operations.

The Disaster Recovery Policies, Procedures and Forms include ten sections consisting of the following:

- DMP100 Introduction
- DMP200 Manual Preparation
- DMP300 Disaster Management Plan
- DMP400 Action Plan
- DMP410 Sample Procedures
- DMP420 Samples Documents
  - Memos
  - Sample Agreements
  - Sample Forms
  - Training Handout
- DMP430 Workplace Violence Guide
- DMP500 Contact Directory
- DMP600 Disaster Recovery Index
- DMP700 Notes

The Disaster Planning Manual manual provides a sample Disaster Management Plan with associated Disaster Recovery Procedures as a guide, but it does not intend to imply these are the "best" or "recommended" text for your specific purposes. They can be used as a "minimum" documentation set for you to use to address generally accepted Recovery Planning practices.

The language style and usage is generally representative of practices in companies in the United States. In some cases information may be used which is not applicable to every business. When you edit and construct a policy it should be easy to read, to the point and convey a message that is clearly understood by both the employee and the management staff. When you have completed your manual, have a team of managers review them and make appropriate comments.

The corrected and finished product is then ready for distribution. But remember, building a manual is not an event; it is a continuous process so revisions will be required from time to time.

> This publication is sold or distributed with the understanding that the publisher is not engaged in rendering legal, IT, or other professional services. If legal advice or other expert assistance is required, the services of a competent professional person should be sought to review the final manual.

Top Ten Management Mistakes

As a busy executive, you face some extremely difficult challenges like creating and dominating new markets or finding and keeping the best people. But then, like many executives, do you find yourself spending too much time solving everyday problems (that only you can solve, right?), which prevent you from growing your ideal business?

You may be making some of these management mistakes:

1. Do you have a compelling vision for your company, that projects a remarkable future, but few of your employees have heard of it or could explain it if asked?
2. Do you have a company mission that addresses your customer needs yet your operations fail to measure your progress towards your mission?
3. Do your objectives focus on increasing revenue and profitability while your assets are performing poorly, generating negative cash flows, or encumbered by debt to create the profit?
4. Do you talk a lot about your employees (positive or negative) without noting what your employee turnover or performance metrics are for your industry?
5. Do you spend a lot of time working IN your business on tactics yet fail to spend a greater amount of time working ON your business to define your strategy, performance metrics, and real resource needs?
6. Do you have regular interactions with employees yet fail to communicate the status of objectives, financials, or metrics?
7. Do you make money available for training yet fail to measure how that training helps your company achieve its goals?
8. Do you constantly strive to improve your company's performance yet fail to compare your performance against external benchmarks for success?
9. Do you believe that your customers, employees, and vendors all love your company yet you have no process for measuring their satisfaction on an on-going basis?
10. Do you produce forecasts and budgets yet fail to achieve the agreed upon goals or learn from the experience to improve in the future.

Daily operational issues eat up much of a manager's time. Too much for most managers. But now, with the purchase of this manual, you have the opportunity to correct all that and build a superior organization that keeps your best people, increases revenue and increases margins.

You now have an opportunity to provide your organization with the methods, tools, and training to achieve superior results. Bizmanualz has helped companies from all over the world build superior organizations with increased compliance, control, and customer satisfaction.

Manual Development Process

The manual development process consists of four phases: Discovery, Planning, Development, and Implementation. W. Edwards Deming PhD, one of the leading experts on process improvement, calls this the Plan, Do, Check, Act (PDCA) cycle. Deming's PDCA cycle identifies the four main phases of continuous process improvement. You should develop your manual in this manner to avoid mistakes that will be made later by skipping these steps.

Many people incorrectly believe that you can save a lot of money by eliminating Discovery and Planning and going directly into Development. But, if you start writing procedures without proper Discovery and Planning then you are in trouble. You can expect constant delays, cost overruns, scope creep, and a lot of rewrites once the missing elements are discovered.

This eventually leads to the feeling that nobody wants to work on this project anymore due to the lack of organization and increasing frustration. But, if you spend the time to plan it well, then the development and implementation phases run a lot smoother. What you are trying to do is anticipate the future.

Discovery

Discovery starts by clearly identifying the project vision, mission, objectives and the action plan. It is considered part of the "Plan phase" in Deming's PDCA cycle. In Discovery you are probably interpreting the results of a previous "Check" phase of some kind. You must have found the results to be dissatisfactory and therefore you are now "Acting" on those results and preparing to "Plan" for the future.

You should include clear business objectives that include how customer needs are getting satisfied with agreed upon effectiveness criteria beyond compliance. Otherwise, the procedures look like overhead and an administrative waste of time to everyone else, making it more difficult to get their buy-in and participation later on during implementation and training.

Once everyone agrees to the objectives and action plan then you can start management planning. The Discover phase takes about 2-4 weeks depending on how organized the executive management team is regarding mission, vision and objectives.

Does your project need a vision?

You bet. Every project requires a vision of the future. All the employees involved will want to know how all of this work will improve their lives. You will need to communicate a picture of "what's in it for them." If you can't explain it to them then all you have is a big wish that it turns out the way you planned. And it won't.

Planning

Planning starts with a gap analysis to determine the difference from reality to the objectives. A Gap Analysis is an audit of your current processes, capabilities, and constraints to determine the "gaps" that need to be filled.

You use the gap results, along with the objectives and action plans from Discovery to produce the management materials needed to control the project and set the budget/expectations. Other functions include creating a development team, assigning process owners for each process and deploying your effectiveness criteria to each process. This takes 2-4 weeks.

This includes producing:

- project plan (activities, resources, dates, etc.)
- project roles and responsibilities
- organization chart
- change & approval review structure
- status reports
- document control and format
- process maps identifying all process inputs & outputs
- compliance requirements
- training, implementation, testing and audit plans.

You are now 1-2 months into the project and should not have written a single procedure. But, everyone should understand the scope of the project, its relationship to the company objectives, what the future will look like, and what their roles are in development.

Development

Development starts by identifying a group of related processes and completing just that group. Follow the document control format, use the review structure, perform a walk through of the related processes, test for compliance and effectiveness, and then move on to the next group until finished. It represents the "Do phase" in Deming's PDCA cycle.

This takes 2-4 months depending on the number of processes, compliance requirements, resource constraints, and the skill and knowledge of the project managers, leaders, writers and reviewers. If you have not done this before plan on four months.

Development also includes creating all supporting documents such as job descriptions, forms, the training plan, and the collection of technical manuals, references, etc, for document & record control and later implementation.

As a rule of thumb you should plan on the Discovery, Planning and Development phases taking as much as 50% of the total project time but consuming 80% of the project cost. You should try to include the time of all personnel working on the project in your project plan so you have a real picture of the costs involved. Frequently it is a lot more than you imagine at first.

You are now 3-6 months into the project and should have completed the procedure writing. You are now ready for a complete document review. You should not implement a smaller group of processes until all process are documented and reviewed as a whole system. Otherwise, you will find yourself rewriting and re-implementing and it will take longer to complete development. You will have a chance to improve the processes later.

Focus on continuous process improvement over delayed perfection.

Implementation

Implementation includes skills assessment, training to improve competencies, and auditing of the entire system against the objectives and compliance requirements. It represents the "Check phase" in Deming's PDCA cycle. The length of time depends on number of employees, locations, and processes. It should take 3-6 months or about 50% of the project time to complete. You are working toward a stable system of processes and this takes time.

Training

Implementation starts with an assessment of employee skills and competencies to determine training gaps. Then training begins introducing the job descriptions, processes, procedures, and the relationship of people/tasks to objectives/effectiveness.

Once training is complete, you should use execute the processes as documented for 1-3 months. Procedure forms, logs, and other records should be used to collect data for measurement and analysis. All personnel should become accustomed to using the processes recently in created. In fact it is the collection of data which now becomes the most important part.

Collected data is used to:

- determine if processes are in control,
- identifying if a process can be improved, and
- demonstrating to auditors and management that the organization is progressing towards achieving the objectives.

Auditing

Auditing is one of the most important steps in building business processes. Auditing provides management visibility into the ongoing progress towards objectives and the capabilities of the organization. Without auditing, management limits its predictive knowledge of the systems potential.

Management Review

You are now 6-12 months into the project and almost finished. The last step is to analyze what's been done, review the lessons learned and look for ways to improve performance, compliance, or effectiveness. This is typically done within a Management Review meeting held after the audit cycle has been completed.

Management holds the review meeting to discuss:

- The audit report documenting the findings of the audit team.
- Progress towards objectives.

- Results of customer satisfaction processes.
- Corrective actions to be taken to resolve common cause errors.
- Preventive action to be taken to resolve special cause errors.
- Follow-up from previous meetings.

Re-Discovery

The actions from the Management Review meeting lead into Re-Discovery or the Discovery process for the next cycle. Re-Discovery starts the process of continuous improvement all over again. It represents the "Act phase" in Deming's PDCA cycle. The result is a new set of objectives and action plans for the next period and you now begin again. This phase overlaps with the discovery phase of the next year to chain together each period in a constant cause-effect cycle of improvement.

Using these four phases you should complete the project in 6-12 months. Less than six months or more than 12 months indicates problems in scope (planning). If the organization is big and complex then the problem should be broken into smaller projects and focused on core processes or, outside contractors should be used to deploy the experience and bandwidth required to meet your schedule.

Staffing for this depends on number of employees, locations, compliance requirements and processes involved. You should consider at least one full-time project leader that drives the project, manages document controls, leads the audits and reports to management on the overall effectiveness.

The project leader could be the Quality Manager. The Quality Manager should be experienced in quality control methodologies, auditing, and the compliance standard you are targeting. If your quality manager is not experienced in these areas then you should look for training. Training is available to learn how to use quality tools, auditing skills, and leading audit teams & programs.

Others are needed to create documents or perform audits. Obviously, there are a lot of variables to staffing requirements. It is difficult for one person to know everything about your business, build the system and then be responsible for auditing the processes. A solo person will lose their objectivity and they can not audit their own processes either.

SECTION 2
YOUR RECOVERY MANUAL

This first step in building your Disaster Management Plan is the creation of an "Action Plan". This is a separate and distinct step from developing your Disaster Management Plan. The purpose is to state, in a concise and brief format, the philosophy, structure, notification procedures, risk analysis, policies and objectives of your company that will be required to achieve your desired level of completeness for your organization or division.

More than likely the input for the Action Plan will come from your auditors, internal and external staff. At a minimum, your Disaster Management Plan must address each one of the business processes of your company and or your industry specific sector requirements.

Each area that you write should cover, at a minimum, three parts: Scope, Policy and Responsibilities. These three areas are used again for each procedure.

- The **Scope** portion describes the areas, functions, individuals or departments affected by the procedure and in what situations.
- The **Policy** portion should state the company policy regarding the applicable Accounting practice.
- The **Responsibility** portion should state who, in generic titles or positions, is responsible for the policy.

A sample Disaster Management Plan is provided in this guide for your use as a template to create your own Disaster Management Plan.

Style and Format:

1. Use a cover or title page.
2. Include a table of contents.
3. Put policy statements on a 8 1/2" x 11 " page and print only on one side to make revisions easier.
4. Organize material by major headings for easy reference.
5. Include an alphabetized index if your manual is lengthy.
6. Avoid a detailed paragraph identification systems of numbers and letters as this will detract from its' readability and the message will be lost.

7. Write in simple easy to understand statements to avoid confusion.

Some companies include sample administrative forms with their procedures along with instructions for their completion. This product includes forms for use at your discretion.

Considerations in writing your manual:

1. It is now common practice to use pronouns that are applicable to either sex or to use his or her, or the more personal and direct, "you". Social changes influence policies on topics concerning smoking, physical fitness, etc.

2. Have your manual reviewed by an attorney to ensure that you are in compliance with all applicable federal, state, and local laws.

3. Define terms used in your manual. Definitions should be placed in each procedure as needed.

Every organization is dynamic and in some state of change. This will lead to changes to your policies from time to time. Revisions should be completed and sent to all personnel who hold a copy of the manual. The revision should have an effective date and of course should be distributed in advance of the effective date. When making a change to your manual be cognizant that the language might have an indirect impact on other policies. And finally, make sure that there is a clear record of revisions made and that all employees have current information in a timely manner.

The Disaster Management Plan as well as each procedure includes a revision section at the end. It is important to keep this revision section up-to-date. It will be the only way to ensure that distributed copies of your manual are current.

Some companies, due to their size and specific business application, require expanded information. For this reason, Bizmanualz.com, Inc has additional business publications that include detailed, topic specific manuals.

The company's Disaster Management Plan is intended to be a simple, "top-to-bottom" guide, addressing the minimum set of statements for the administration and creation of a workable Disaster Management Plan. Most guidelines presented have immediate use and importance. However, by using the MS-WORD documents you can edit the samples and customize them to your company's individual needs.

Bizmanualz.com, Inc is committed to providing professional publications for those business owners and managers dedicated to the development and success of their company. To this end, we will continue to publish useful business guides to assist you in your ever-demanding endeavors. To obtain the latest information on all of our products, visit our website at www.Bizmanualz.com .

Sources of additional Information

With the help of the prewritten documentation contained in this publication, you should be able to produce an effective Disaster Management Plan. However, you may be able to draw on other sources of information to develop a comprehensive program that truly meets the needs of your organization.

These sources can include:

1. Industry or trade association publications
2. Industry or Sector Guides
3. The company's legal counsel
4. Other related company Manuals and Procedures
5. Internal memos and records
6. Equipment User Manuals
7. Regulatory agencies
8. Standards Organizations Small Business Advisory Centers

[This page intentionally left blank]

SECTION 3
EFFECTIVE COMMUNICATION

Communication and Addressing Your Audience

It is vital in order for your manual to be effective that it be clearly written and easily understood by all the employees in your organization. Remember that the objective of your Disaster Management Plan is to improve you're ability to effective resume business operations of your company after some type of disaster. Therefore, the most important rule when writing a Disaster Management Plan that clarity and completeness are much more important than style, perfect grammar and a large vocabulary.

The writer must try to put themselves into the position of the users at all times. While developing policy and procedure statements, some general guidelines for the writer to keep in mind are:

1. Explain new or unusual terms the first time they are used or in the definitions section of the statement.
2. Avoid jargon wherever possible, especially when training new employees.
3. Avoid unneeded words.
4. Avoid complex writing. If the writer's vocabulary is unusually large or if they write using complex sentence structures, the writing may be at too high a level for many of the users. Understanding is far more important than the correctness of the language.
5. Use active verbs instead of passive phrases.
6. Write the way you speak. Use words and phrases that you would normally use in expressing the same thought orally.

Sexism in Writing

Webster's New Collegiate Dictionary defines "he" as used in the generic sense or when the sex of the person is unspecified. However, many people will not accept "he" being used when referring to people in general. There have been suggestions that the generic "he" be replaced with "he or she" and "him" with "him or her."

Another recent method that is gaining acceptance is the switch to a plural pronoun with a singular subject such as "When someone orders their supplies, they will have to complete a ..."

Often it is more practical for the writer to use generic nouns or by recasting sentences to include positions or titles (i.e., Applicant, Manager, Accountant, Driver, etc.) to eliminate the need for most sexism in writing of policy and procedure statements.

Number Usage

Writing and development of procedures may often times include the frequent usage of numbers in the writing. The following rules cover how numbers should be used in print.

1. Never begin a sentence with numbers. "Fifty states have been admitted to the union" not "50 states ..."
2. Spell numbers one through nine in words, 10 or larger in numerals.
3. Compound numbers such as fractions in numerals, for example 8 1/2" X 11."
4. When showing odds, use hyphens such as 4-to-1.
5. Decimals of less than one should be preceded by a zero, for example, 0.5 percent.
6. Dates: "the 1990s"
7. Spell % as percent.
8. Use numerals with times such as a.m. and p.m.

Organizing Your Thoughts

To help the writer formulate and organize ideas for developing and writing policy and procedure statements, it may be useful to outline the material to be covered. Outlining is a fast and effective way to show a great amount of information in a concise, efficient manner with a minimum of writing.

To achieve well-written and easily understood policy and procedure statements that flow in a cohesive and logical form, your personnel should first outline their thoughts before beginning to write a statement.

When outlining a policy and procedure statement, the following areas should be defined:

1. What is the objective that the statement is going to accomplish?
2. What is the Company's policy on this matter?
3. Who is affected?
4. When is the policy/procedure appropriate?
5. How is the objective to be accomplished? This should include outlining major areas in a step-by-step fashion in chronological order.

Outlining Technique

There are some basic standard rules for effective and consistent outlining. It may be helpful to briefly review a few of these that deal primarily with formatting. These rules concern indentation and numbering.

Standard outline formatting is as follows:

I.
    A.
        1.
        2.
            a.
            b.
    B.
II.

There should always be at least two of each type of character. For example, there should not be an I without a II, an A without a B, etc.

The four characters to mark each section level, Roman numerals, capital letters, Arabic numerals and lower case letters are followed by a period. Line up the periods vertically for each type of character. Thus, Roman numerals are shown:

 I.

II.

Arabic numerals are lined up as follows:

 5.

39.

By aligning the periods, the text is a pleasing, straight vertical line. Indentation should be uniform, usually two to five spaces for each change in type of line. The part of a single line carried to a second line should be indented the same number of spaces as the original part of the line.

Defining the Format and Organization of Your Manual

After reviewing the preliminary listing of statements to be included in the manual and discussing with the personnel assigned to each section, you should be able to determine the estimated length and usage of your manual. With the guidelines presented in this publication and from your own organization's needs determine the format of your policy and procedure statements and how you will organize the manual into sections.

The format and appearance of your policy and procedure statements are just as important as the organization and content of the manual. A manual that is appealing to the eye and that emphasizes the importance of the statements is more likely to be taken seriously and used on a regular basis by employees.

However, it is important to remember that the true objective of the manual is to disseminate information in a timely and efficient manner and not to "impress" the reader with intricate headings or fancy printing techniques.

The simplest format is often the best. A simple format also allows for the most time and cost effective manner for production and maintenance of the manual. Therefore, it may be best to avoid temptations such as, detailed corporate logos in headings, two sided copies, odd sized paper patterns, expensive and restrictive binding techniques, etc.

Design Features

No Disaster Management Plan should ever be regarded as "complete" in the sense that it will never change. The best manual is one that is geared to continuous growth over time and incorporates design features that make this kind of growth possible.

In this regard, use of standard 8-1/2" X 11" paper housed in a three-ring binder forms an ideal employee handbook manual.

The three-ring binder provides the benefits of allowing a place for statements to be inserted while the manual is being developed and provides for easy updating through simple replacement of pages or superseded statements. Further, as the organization grows, the use of standard three-ring binders allows additional copies of the manual to be produced on an as needed basis instead of having to be concerned with minimum production runs required for hard bound versions.

Production of statements on a single-side, standard size paper medium provides for easy reproduction of the manual by high-speed copiers. Use of single-side printing also provides for easy updating of the manual with changes by allowing for one or two pages to be replaced without affecting the order or sequence of the manual.

However, if the manual becomes too voluminous for ease of handling, it may be necessary to bind the manual by different sections or utilize two-sided printing to reduce paper volume.

A window type binder should be used to describe the contents of the manual on the spine for locating sections quickly from a bookcase. If desired, the outside of the binder can be imprinted with the company's name and logo to give the manual a more professional and authoritative appearance.

Divider tabs on heavy stock should be used to separate functional areas or departmental sections for ease in finding a specific statement.

Style and Mechanics

The style and mechanics of writing include the paper, typestyle and print quality.

Paper - Some organizations use color to designate different manuals, forms, memos, etc. While color can create a pleasing appearance, it sometimes becomes too restrictive and may complicate photocopying or printing. Further, sometimes colors do not provide adequate contrast from the ink color for ease in reading. Nothing is better than black ink on white paper.

The grade of paper is usually not important since the manual is for internal purposes only and is not intended as a public relations effort. Regular 20-pound copy paper is adequate for single-sided printing. A heavier weight paper with greater opacity may be necessary if two-sided printing is used.

Typestyle - Avoid unusual artwork or type styles. These can be difficult to read and/or reproduce over a long period of time. You should strive for consistency in the overall appearance of the entire manual regardless of what area or department originates the statements by selecting a common typewriter element or word processing typestyle font. Courier 10, Elite, 12-point times roman and 12-point roman are common typestyles used by many business machines and word-processing software packages.

Avoid using small print, photocopy reductions, all-capital print or fancy script styles whenever possible as these are tiresome and difficult to read.

Adequate margins should be provided on the page. Recommended margins are a 1" top and bottom margin with a 1" left and right mirror margins plus a 1/2" gutter margin (for three-hole punching).

Sources of additional Information

There are a number of books on the market that provide an excellent reference for technical writers, editors, and others developing policies and procedures.

These sources can include:

1. Microsoft Manual of Style for Technical Publications; ISBN: 1572318902
2. The Handbook of Technical Writing; ISBN: 0312254962
3. Starting a Documentation Group; ISBN: 0967417902
4. Managing Your Documentation Projects; ISBN: 0471590991
5. Writing Effective Policies and Procedures; ISBN: 081447960X
6. Establishing a System of Policies and Procedures; ISBN: 1929065000
7. Achieving 100% Compliance of Policies and Procedures; ISBN: 1929065493

[This page intentionally left blank]

SECTION 4
DISASTER RECOVERY PROCEDURES

Format

Disaster Management Plan Procedures address the primary steps or tasks of an activity for a department or function and the personnel (generic titles or positions) responsible for accomplishing the procedure(s). These procedures can be organized on a departmental basis.

The Disaster Management Plan Procedures in this manual start with a designation such as "DMP" for Disaster Management Plan and may be used as a template to create your own procedures. The exact format is not required but some of the field names and/or titles are accepted as common usage in the Disaster Recovery environment.

Heading Information

Many organizations use very complex headings that require space of up to one-third of a page or more and can include multiple approval signatures, distribution information, structured tabular formats and confusing retrieval numbering systems and revision levels.

The simplest approach is often the best. Heading information should be kept to the minimum necessary to accurately describe the statement, identify the revision level, demonstrate authorization and be easy to produce.

A sample heading may appear as follows:

SAMPLE COMPANY	
SOP # G1 Revision 0	Prepared by:______
Effective Date: 11/01/0X	Approved by:_____

The above heading provides the following information:

Formal Company Name or Division - Use of the company's name personalizes the statement as belonging to "our" company and not just boilerplate borrowed from another company. The company name should be located in the upper left-hand of the page.

SOP Number and Revision Level - For small manuals, no numbering system may be necessary and just the title can be used for identification purposes. With lengthy manuals, it is best to use a simple alphanumeric numbering system for identifying SOPs and for storage and retrieval purposes in data processing systems.

The first digit(s) of the number is an alphabetical character that represents the primary department or functional area covered by the SOP. For example, table5.1 shows how alpha characters could be used to describe these departments:

A	=	Accounting
C	=	Credit
CS	=	Customer Service
E	=	Engineering
F	=	Finance
G	=	General & Administrative
M	=	Manufacturing
MK	=	Marketing
P	=	Personnel
QP	=	Quality
S	=	Sales

Table 5.1 Alpha Department Codes

The remaining digits should be numeric and assigned sequentially as statements are developed and issued. For example, the first Accounting procedure could be A1, and the second one A2, etc. An alternative would be to designate two or three digits to a numeric assignment and begin at a starting level of 11 or 101. For instance, beginning Accounting procedures could be assigned as A101 and A102. Or beginning engineering procedures could be E101 or E102.

To allow for ease in updating and development of the manual it is best to try to avoid grouping similar SOPs together in sequential order. However, if it is desired to group similar procedures together numerically, a decimal numbering system may be used to insert a new statement between two existing SOPs. For example a new SOP could be numbered A1.1 and inserted between A1 and A2.

Once an SOP is issued, it will be subject to change and modification. A revision code should be used to identify the current one in effect and assist with purging obsolete statements. An initial SOP begins with a revision level of zero (0). If a new SOP supersedes this prior statement, the revision number will be increased by an increment of one. The SOP number and Revision level should be located underneath the company name.

Effective Date - The actual date the SOP or Revision will be implemented. Note the effective date is <u>not</u> the issue date of the original SOP. The effective date should be in the format of MM/DD/YY and is located directly beneath the SOP number and Revision level.

Page Numbering - All pages should be numbered in the form Page # of # to quickly identify the order or placement of pages within a statement. It is also useful when updating an existing SOP where only one or two pages will be replaced. The page number should be placed in the upper right corner of the page.

Subsequent pages after the title page should be identified by either the SOP # or the title along with the page number. For example, the second page of the above SOP heading would be as follows:

SOP # A1

Prepared by: - It is useful to identify the primary individual or department that has developed the statement in the event questions arise during the approval process or subsequent to the issuance of the SOP in which additional information may be needed or to be clarified. This identification can be either typed in or simply initialed by the individual. The "Prepared by" identification should be placed beneath the Page Number.

Approved by: - After the SOP has been properly reviewed and authorized, the title page should be initialed by an authorized individual. It is usually best to initial the approval section by hand rather than to type it in to clearly indicate that the SOP has been authorized and distinguish the final version from any draft versions that may still be in circulation. The "Approved by" identification should be placed directly underneath the "Prepared by" section.

Title And Introduction

The title and introduction section provides the information necessary for the reader to determine content and the reason for the SOP, whom it affects and in what situation(s), the Company policy covering this area and the definition of any new or unusual terms.

A sample Title and Introduction section can appear as follows:

Title: DEVELOPMENT AND MAINTENANCE OF STANDARD OPERATING POLICIES AND PROCEDURES

Policy: It is the policy of Sample Company that all departments will prepare and maintain standardized operating policies and procedures that cover the performance of all major functions within their department.

Purpose: This procedure outlines the steps involved in preparing, maintaining and approving standard operating policies and procedures in order to provide consistent, informative and effective procedures to the employees of Sample Company.

Scope: This procedure applies to all policies and procedures used or written by all departments and individuals of Sample Company.

Responsibilities:

All Personnel are required to understand and use this procedure.

Management is responsible for maintaining this procedure.

Definitions: Policy - A definite course or method of action to guide and determine present and future decisions. It is a guide to decision making under a given set of circumstances within the framework of corporate objectives, goals and management philosophies.

Procedure - A particular way of accomplishing something, an established way of doing things, a series of steps followed in a definite regular order. It ensures the consistent and repetitive approach to actions.

A general description of the Title and Introduction is as follows:

Title - The title of the SOP should be short but descriptive enough to allow the reader to determine the content or objective of the SOP.

Policy - The policy should clearly indicate the company's or top management's beliefs or protocol affecting this area.

Purpose - A brief description of the objective of the SOP. Similar to Policy.

Scope - Describes the areas, functions, individuals or departments affected by the SOP and in what situations.

Responsibilities - Describe who, in generic titles or positions, is responsible for implementing or maintaining the procedure or parts of the procedure.

Definitions - Describes any terms contained in the SOP that may be new or unusual to the reader.

The Body Of The Procedure Statement

The body of the SOP includes a complete description of the policy and/or the procedure, the methods to be used, form names, cross references to other SOPs related thereto, etc.

Although the narrative information in the body can vary considerably in format, it is imperative that it clearly explains to the reader in an orderly fashion exactly how to accomplish the objective of the procedure.

For lengthy statements, it may be useful to identify and segregate the steps or areas by a numbering system. For example, each procedural category would be identified with a step number, starting with 1.0 and a heading description. All the steps within the category would then be numbered sequentially (For example, 1.1, 1.2, 1.3, etc.). The step number should be located at the left margin with the related narrative indented.

A sample procedure section can appear as follows:

Procedure - Describe the main elements that make up the procedure. Include the steps or details, the methods to be used, form names, cross-references to other SOPs related thereto, etc.

Effectiveness Criteria - Describe any thresholds or standards that are used to evaluate the work product or results of the procedure. How will an employee know that the procedure was executed correctly?

References - List applicable documents, procedures, manuals, laws, regulations or other sources that were used to develop or influence the policy/procedure statement.

Records - Describe the records, minutes, reports, notes, forms, or other documents that are generated or used when implementation the procedure.

Revision History - Describe all revisions made to the procedure. Include a revision number, date of the revision, a short description of the changes made, and a source of the request for the change.

Attachments

Any forms, diagrams, illustrations or other documents referenced in the body of the statement should be attached to the SOP and referenced as exhibits in the Records section using a sequential numbering system, (For example, Exhibit 1, Exhibit 2, etc.). It may be useful to use copies of actual completed forms or documents as exhibits to illustrate to the reader how they are completed.

Once all comments have been received and a final version approved, the statement should be printed in its final form. The statement should then be authorized by the appropriate individual and released for production and distribution.

Authorization

Origination of policies and procedures usually begins at the unit level by employees or department managers. Once a draft copy of a proposed procedure is developed it should be reviewed, corrected if necessary and approved before being released as a corporate policy and procedure statement. The approval process generally consists of review for consistency and accuracy, conflict with corporate policy or with other statements and general readability.

The approval process can vary widely between companies but it is recommended to keep approvals to a minimum if possible. If too many people or managers of equal ranking are required to authorize a statement, it can turn the development of statements into a bureaucratic process, which can considerably slow the release of policies and procedures while adding little value to the final version.

A method for gaining the input of others while streamlining the process and keeping authority at the functional or departmental level is to release draft copies of proposed statements to a select number of individuals for comment. It should be made clear to these individuals that they should confine their suggestions to what they feel is really essential to the statement's accuracy, readability and usefulness.

The selection of who should receive a draft copy will depend primarily on the nature and content of the statement. Sensitive issues or areas that deal with corporate exposure such as in the personnel area, intellectual property or trade secrets should be reviewed by top management including the president and may also include the company's legal counsel.

Rudimentary procedures that affect only a small unit within the company and are likely to be of no interest to others should be kept to a minimal review process. However, it is still advisable to have someone familiar with the area but separate or outside of the unit or department review the proposed statement. For example, the company's finance officer may review a proposed accounting SOP. This type of review serves three purposes:

First, what makes sense to the preparer directly involved in enforcing the policy or conformance to a procedure may not be understood when read for the first time by someone not as closely involved.

Second, a review by multiple department managers may prevent a later conflict with a policy and procedure statement still in the formative or discussion stages elsewhere in the company of which the original preparer would have no knowledge.

Third, it allows the input of multiple individuals while allowing the department manager to maintain control of the integrity of the statement and to drive its completion and release in a timely manner.

Production And Distribution

Once a policy and procedure statement is authorized, it should be duplicated on standard white copy paper and three-hole punched on the left margin. Multiple page statements may be corner stapled to prevent losing pages until it is included in the manual.

The number of copies of the statement should correlate to the number of manuals that have been distributed. It is generally advisable to designate one individual in the company for the production of statements. This individual keeps track of the number of manuals issued and ensures that new or revised SOPs are distributed to the appropriate personnel.

Management of the company should decide which departments or positions will receive copies of the manual or as an alternative, sections of the manual may be distributed which pertain only to a specific function or department. However, if the manual is to serve as a communication tool, enough copies of the manual should be available to employees.

Since the manual contains many operating procedures that are vital to the company's business practices and methods, there should be some accountability for the manuals. Generally, one individual will maintain a list of the number of copies in circulation and the names of those to whom they have been assigned. When a supervisor or manager leaves the company, there should be a strong incentive to return their copy of the manual; some companies withhold final compensation until the manual has been returned.

However, one should avoid numbering each copy of a manual, unless absolutely necessary. Numbering implies confidentiality and some degree of importance, which may not be the case. Besides, the issuer must maintain a permanent record of the numbers cross-referenced to the recipients that can make personnel changes a tedious record-keeping task.

Revising And Updating Statements

As mentioned previously, the policies and procedures manual is never "complete" because it never stops changing or evolving in order to keep pace with needs of the business and business climate.

All employees should be encouraged to initiate changes or revisions to existing policy and procedure statements that affect their area of responsibility. This will assist the company greatly in keeping the manual current and up-to-date.

In addition to this continuing review process, the entire manual should undergo a complete audit as often as changing conditions both within and outside the company indicate. This may be every year or several years depending on the company's business environment.

A new SOP should be issued if an existing statement is to be modified in any way. The revised SOP should undergo the same approval process as the initial statement and should be assigned a new revision number level to indicate that it supersedes the prior statement. Superseded statements should be purged from the manual immediately and discarded.

[This page intentionally left blank]

<COMPANY>

DISASTER RECOVERY PLAN 300

[This page intentionally left blank]

DISASTER RECOVERY PLAN

Table of Contents*

* The Disaster Recovery Plan Table of Contents references section page numbers (lower right corner.

[This page intentionally left blank]

1.0 POLICY AND PROCEDURE STATEMENTS

1.1 Purpose

The purpose of this section is to define and to explain the format and operation of the company's Disaster Management Plan. All statements issued according to this procedure shall be applicable to all personnel of the company.

1.2 Definitions of Statements

POLICY STATEMENTS - A policy statement is one describing a policy or program of broad application throughout the company, and will be used as a guiding document until it is withdrawn or modified.

PROCEDURE STATEMENTS - A procedure statement is one describing the manner in which a policy is to be implemented, and may include performance expectations and specific tasks to be performed.

MEMORANDA - A memorandum is a document describing an informal request or notification not in conflict with existing policies or procedures, unless used as an interim emergency measure.

1.3 Issuance of Statements

POLICY STATEMENTS - must have a basis of both legality and authority. There should be a supporting legal provision based on state law, personnel rules and regulations, or other lawful enactment. The person issuing the order must be the one to whom appropriate authority has been delegated for its issuance.

PROCEDURE STATEMENTS - must have a basis of both sound research and applicability. No procedure statement should be issued without reference to or implementation of an accompanying statement of policy or purpose.

No authority is needed for the issuance of MEMORANDA, which shall contain the name of the author, intended recipient, and a statement of purpose. A copy of any memorandum issued by a member of the Disaster Management Team, and affecting the plan's implementation, shall be forwarded to the Disaster Management Team Chairpersons and Coordinators for review before distribution.

1.4 Distribution of statements

Copies of the Disaster Management Plan are available for review at specific locations within the company as follows:

1. Disaster Management Team Chairpersons (full copy).
2. Disaster Management Team Coordinators (full copy).
3. Each Office Manager (limited copy: position-specific).

4. Each Department Manager (limited copy: department-specific).
5. Centers of Operations (full copy).

1.5 Maintaining Disaster Management Plan Manuals

The Disaster Management Team Coordinators shall coordinate the reproduction, placement and updating of all copies of the company's plan. The Coordinators will be responsible for appropriate maintenance of the Disaster Management Plan manuals, to include:

1. Updating of all copies of the manuals when changes or revisions are made to any statement.
2. The assignment of section numbers and other appropriate information.
3. Conducting at least an annual review of all statements. This shall be conducted in the same month of each year, to determine the needs for cancellations or revisions. After this review is completed, a summary will be presented to the Disaster Management Team Chairpersons regarding new, updated or outdated information.
4. Furnishing copies of all changes for each location.

1.6 Cancellation and Purging

1. POLICY STATEMENTS - Policy Statements are not self-canceling, and shall remain in effect until canceled or superseded.
2. PROCEDURE STATEMENTS - Procedure Statements indicating an indefinite date of effectiveness will be reviewed in the same manner as Policy Statements.
3. MEMORANDA - N/A.

1.7 Index

1. The Disaster Management Plan shall contain at least a table of contents, and an index or appendix, if appropriate.
2. Indices will be periodically reviewed and, when sufficient changes have been made, new indices will be prepared and published.

2.0 INTRODUCTION & PURPOSE OF DISASTER MANAGEMENT PLAN

2.1 The Disaster Management Plan is designed to assist the company in effectively planning for, and acting during, emergencies that might otherwise impact the company's operations.

2.2 The continued research, development and implementation of the company's Disaster Management Plan involves the participation and approval of each department within the company.

2.3 The Disaster Management Plan addresses these topic areas, in order of priority:

1. Personnel
2. Customers
3. Facilities
4. Assets
5. Records

2.4 For the purposes of this plan, a disaster is defined as any event or circumstance that may significantly impact the company's operations. This includes any sudden, unexpected occurrence, natural or human-caused, which poses a significant threat to the company's personnel, customers, facilities, assets, records, and its delivery of services.

2.5 The purposes for developing and implementing this Disaster Management Plan are to:

1. Determine the critical and accessory functions of the company. This involves a continuing assessment of the company's:
   - Assets and liabilities
   - Personnel
   - Equipment and supplies
   - Facilities
   - Policies and procedures
   - Legal obligations
   - Existing systems
   - Organizational structure
2. Develop and implement a written recovery plan based upon critical and accessory functions for each office, facility, department and function within the company.
3. Continually assess the risk of loss of support to each function and department within the company.
4. Continually identify appropriate solutions to reduce any projected impact upon each function and department. This involves the cooperation with, and delegation to, members within each office, function and department.
5. At least annually reassess primary and secondary solutions based upon anticipated or demonstrated results. After each re-assessment, company managers within each function and department shall assign the responsibility for the actual implementation of the Disaster Management Plan.
6. This assignment shall be to those personnel whose individual areas of responsibility are affected.
7. Test the results at least annually, in a controlled environment, for validity.
8. Either retain those procedures or propose new ones in writing to the Disaster Management Team Coordinators.

3.0 PLAN DEVELOPMENT & IMPLEMENTATION: GUIDELINES & SCOPE

3.1 Guide Lines

In conjunction with the stated purposes and objectives, the following guidelines shall be in effect:

1. Effective plans shall be continually developed and maintained for each department, facility and support unit. These plans shall be in line with the critical and accessory functions to be fulfilled by that department, facility or support unit.
2. Each department, facility and support unit shall assist in the continual development and implementation of policies and procedures created during this process, and shall be responsible for achieving the necessary results.
3. All policies and procedures shall be written in a clearly defined and carefully constructed manner, and itemize specific objectives and concepts.
4. All policies and procedures shall be designed according to appropriate corporate (if applicable), contemporary business, and legal standards. These policies and procedures shall also be fully supportive of other company plans, goals and objectives.
5. This Disaster Management Plan shall not be used as a substitute for reasonable Preventive measures now in place in any department, facility or support unit.
6. The acquisition of any new or remodeled equipment, devices, systems or procedures shall be subject to examination before purchase. This is to determine the extent of support that such acquisition will provide to this plan. These acquisitions shall support this plan to the fullest extent.

3.2 Essential Functions

These functions are essential to the company's operations and, in the event of a disaster, should be operational within **8-24 hours**:

1. Administration.
2. Telecommunications.
3. Contact with the Board of Directors.
4. Public relations (critical functions only).

3.3 Critical Functions

These functions are critical to the company's operations and, in the event of a disaster, should be operational within **24-48 hours**:

1. Security.
2. Personnel services (critical functions only).
3. Office and customer service operations; (critical functions only).

4. Delivery, purchasing and courier services.
5. Credit and accounting services (critical functions only).
6. Office support (critical functions only).
7. Data processing (critical functions only).

3.4 Supplemental Functions

These functions are supplemental to the company's operations and, in the event of a disaster, will be restored as appropriate upon the resumption of the company's critical functions:

- Insurance and legal
- Marketing and public relations
- Accounting and bookkeeping
- Purchasing, delivery and receiving
- Audit
- Accessory functions of other named function

4.0 DISASTER MANAGEMENT TEAM

The following personnel comprise the Disaster Management Team:

CO-CHAIRPERSON: ______________________________

Responsible for:

- Board of Directors
- Public and Media Relations
- Shareholder and Stockholder Relations
- Administration

Backup:

CO-CHAIRPERSON: ______________________________

Responsible for:

- Board of Directors
- Public and Media Relations
- Administration
- Personnel
- Insurance
- Accounting
- Data Processing, Administrative
- Functions

Backup:

CO-COORDINATOR: ______________________________

Responsible for:

- Operations
- Security
- Telecommunications
- Data Processing, Operations
- Functions

Backup:

CO-COORDINATOR: ______________________________

Responsible for:

- Operations
- Courier Services/Transportation
- Telecommunications
- Office Support
- Delivery/Purchasing, Administrative Functions

Backup:

MEMBER: ______________________________

Responsible for:

- Credit Administration
- Courier Services/Transportation
- Delivery/Purchasing,
- Operations Functions.

Backup:

MEMBER: ______________________________

Responsible for:

- Data Processing

Backup:

MEMBER: ______________________________

Responsible for:

- Personnel Services and Payroll
- Accounting and bookkeeping

Backup:

MEMBER: ______________________________

Responsible for:

- Customer Service
- Backup:

MEMBER: ______________________________

Responsible for:

- Accounting/Bookkeeping
- Backup:

MEMBER:

____

Responsible for:

- Audit Compliance
- Legal
- Backup:

MEMBER:

____

Responsible for:

- Facilities
- Equipment and Supplies
- Transportation
- Backup

MEMBER: ______________________________

Responsible for:

- Disaster Records
- Backup:

MEMBER: ______________________________

Responsible for:

- Office Operations #1
- Backup

MEMBER: ____

Responsible for:

- Office Operations #2
- Backup:

MEMBER: ____

Responsible for:

- Office Operations #3
- Backup:

5.0 NOTIFICATION OF EMERGENCY RESPONSIBILITIES

5.1 Each officer selected to perform emergency responsibilities as a member of the Disaster Management Executive Committee, shall be notified of his/her selection, and of the duties and responsibilities that each person is expected to perform.

This selection may be declined without risk.

5.2 Each person accepting these responsibilities shall attend meetings and training events concerning the Disaster Management Plan, and shall remain up-to-date regarding the policies and procedures developed as a result of these meetings.

5.3 To provide for the necessary redundancy in appointments, the positions of Chairperson, Coordinator and Department Leader shall be appointed as a team of at least two persons.

A designated backup person shall be appointed for all people assigned critical emergency duties and responsibilities.

6.0 DUTIES AND RESPONSIBILITIES

6.1 Disaster Management Team

The duties and responsibilities of the Disaster Management Team shall include all phases of the Disaster Management Plan:

1. Delegate the initial and continuing research and development of specific elements of this plan to appropriate personnel, as those elements pertain to their departments and/or assignments.
2. Ensure that appropriate guidelines are provided for this initial and continuing research, development and implementation of this plan.
3. Review the information gathered, and create or update each office, departmental and function plan for inclusion into the total plan.

4. Arrange for the initial and continuing implementation and testing procedures involved with all phases of this plan.
5. At the direction of the Disaster Management Team Chairpersons, implement appropriate recovery, restoration, resumption and reconstruction efforts as applies to their specific areas of responsibility, or other areas as assigned by the Chairpersons.
6. Act as liaison with law enforcement, fire and medical personnel, and provide the Disaster Management Team Chairpersons with timely and complete updated information regarding any emergency addressed by this plan.
7. Ensure that the disaster management activity logs are prepared by all departments, indicating events and measures taken to address the impact of the emergency.
8. Personally ensure that their specific areas of responsibility are staffed with appropriate personnel, and that these personnel are working in a safe environment, to the degree possible.
9. Coordinate the post-disaster evaluation effort, and advise the Chairpersons of the results by means of a written report.
10. Participate in the company's periodic overall evaluation effort, and provide written recommendations regarding changes to be made to the Disaster Management Plan.
11. Schedule and hold at least annual meetings with all Team members, and ensure that each meeting is documented regarding:
    1. The purpose of the meeting.
    2. Those persons in attendance.
    3. Agenda items addressed.
    4. Revisions proposed during the meeting, to be set forth in an interim policy format, to be discussed and approved during the next regularly-scheduled Board of Directors' meeting.

6.2 Disaster Management Team Chairpersons

The Disaster Management Team Chairpersons shall be appointed by the company's Board of Directors, and shall be responsible for the following duties and responsibilities:

1. Ensure the timely, continual and effective research, development and implementation of the Disaster Management Plan.
2. Develop, assign and monitor the duties and responsibilities of the other Team members.

3. Review the results of each developmental, implementation and testing phase of the plan when appropriate, and arrange for the plan's annual certification by the company's Board of Directors.
4. Maintain a current master copy of the plan secured at his/her primary work place, and a duplicate master copy at his/her place of residence or other appropriate location.
5. Upon notification of a potential or current emergency, determine the potential and/or current damage to facilities and operations.
6. Verify that all appropriate law enforcement, fire and medical services have been notified of the emergency.
7. Ensure that all Disaster Management Team members have been notified of the emergency, either in person or by phone, and that those members are responding immediately to the Disaster Management Center of Operations.
8. Determine the cause of the emergency, if possible.
9. Ensure the execution of necessary measures to provide for the safety of personnel, customers, facilities, assets and records.
10. Initiate the shift in operations to the designated Disaster Management Center of Operations, if appropriate.
11. Ensure that all recovery, restoration, resumption and reconstruction phases of this plan are conducted appropriately.
12. Provide the Board of Directors and other appropriate personnel with updated information on a timely basis, regarding this emergency and its impact.
13. Upon completion of the emergency, ensure that all measures taken by the company's personnel during the emergency are recorded.
14. Present a synopsis of the results to the Board of Directors for evaluation.
15. Ensure that the company's Disaster Management Plan was implemented as appropriate, and/or arrange for the plan to be modified based upon the results of this emergency.

6.3 Disaster Management Team Coordinators

The Disaster Management Team Coordinators shall be appointed by the company's Board of Directors, and shall be responsible for the following duties and responsibilities:

1. Research, develop, implement and continually monitor a comprehensive emergency response and disaster recovery policy and procedure mechanism. This is to integrate existing precautionary programs with programs to be developed.

2. Maintain files for each department and facility regarding emergency response and recovery measures to be performed by that department or facility.
3. Research, develop, implement and monitor an emergency evacuation plan for each facility, and submit each plan and at least annual evaluations to the Disaster Management Team for review.
4. Designate and train emergency leaders for all offices, departments, and facilities, to assume command during emergency responses.
5. Research, develop, implement and monitor a comprehensive emergency response and disaster recovery training program for all personnel.
6. Maintain a current master copy of the plan secured at his/her primary workplace, and a duplicate master copy at his/her place of residence or other appropriate location.
7. Arrange for or supervise the acquisition of all sites, equipment, supplies and services necessary for both the primary Disaster Management Plan Center of Operations and all alternate sites.
8. Manage and control all aspects of emergency activation and disaster response planning, testing, implementation and maintenance of systems, procedures and policies.
9. Collect, analyze and report upon facts concerning all departmental functions relative to emergency activation and disaster response.
10. Conduct at least annual reviews and assessments of the protective measures designed to reduce risks.
11. At the direction of the Disaster Management Team manage, direct and control assigned aspects of the recovery portion of this plan, including the assignment of personnel and the allocation of resources.
12. At the direction of the Disaster Management Team, implement restoration procedures and conversion to normal operations at the termination of the emergency.
13. Provide the Disaster Management Team Chairpersons and other appropriate personnel with updated information on a timely basis, regarding this emergency and its impact.
14. Upon completion of the emergency, ensure that all measures taken by the company's personnel during the emergency are recorded, and present the recorded results to the Disaster Management Team Chairpersons for evaluation.

6.4 Disaster Management Team Department Leaders

The Disaster Management Team Coordinators shall appoint Department Leaders (Office Managers shall function as Department Leaders), who shall be responsible for the following duties and responsibilities relating to the Disaster Management Plan:

1. With the assistance of the Disaster Management Team Coordinators, research, develop, implement and continually monitor a comprehensive emergency response and disaster recovery policy and procedure mechanism. This is to integrate existing precautionary programs with programs to be developed.
2. Maintain files for his/her department and facility regarding emergency response and recovery measures to be performed by that department or facility.
3. With the assistance of the Disaster Management Team Coordinators, research, develop, implement and monitor an emergency evacuation plan for his/her department or facility, and submit each plan and at least annual evaluations to the Disaster Management Team Coordinators for review.
4. With the assistance of the Disaster Management Team Coordinators, designate and train emergency leaders and assistants for all his/her department or facility, to assume responsibilities during emergency responses.
5. Research, develop, implement and monitor a comprehensive emergency response and disaster recovery training program for all personnel, according to the guidelines as supplied by the Disaster Management Team Coordinators.
6. Maintain a current master copy of the plan secured at his/her primary work place, and a duplicate master copy at his/her place of residence or other appropriate location.
7. Recommend to the Disaster Management Team Coordinators the acquisition of all sites, equipment, supplies and services necessary for his/her department or facility to respond adequately to an emergency or disaster.
8. With the guidance of the Disaster Management Team Coordinators, manage and control all aspects of emergency activation and disaster response planning, testing, implementation and maintenance of systems, procedures and policies for his/her department or facility.
9. Collect, analyze and report upon facts concerning all departmental functions relative to emergency activation and disaster response for his/her department or facility.
10. Conduct at least annual reviews and assessments of the protective measures designed to reduce risks, and furnish those written results to the Disaster Management Team Coordinators.

11. At the direction of the Disaster Management Team Coordinators, manage, direct and control assigned aspects of the recovery portion of this plan, including the assignment of personnel and the allocation of resources.
12. At the direction of the Disaster Management Team Coordinators, implement restoration procedures and conversion to normal operations at the termination of the emergency.
13. Provide the Disaster Management Team Coordinators and other appropriate personnel with updated information on a timely basis, regarding this emergency and its impact.
14. Upon completion of the emergency, ensure that all measures taken by the company's personnel during the emergency are recorded, and present the recorded results to the Disaster Management Team Coordinators for evaluation.

6.5 Disaster Management Staff Personnel

The Disaster Management Team Department Leaders shall appoint appropriate staff personnel, to be responsible for the following duties and responsibilities relating to the Disaster Management Plan:

1. At the direction of the Department Leaders, participate in the research, development, and implementation of a comprehensive emergency response and disaster recovery policy and procedure mechanism for his/her department.
2. With the guidance of the Department Leaders, to assume appropriate responsibilities during emergency situations, and to provide for the implementation of appropriate disaster recovery procedures as relates to his/her position within the company.
3. Guidelines for the performance of emergency tasks shall be developed for each position within the department or facility. These guidelines shall be placed at the end of this section, and this information shall be updated at least annually, or whenever a position is added or changed significantly.
4. Participate in a comprehensive emergency response and disaster recovery training program for all personnel, according to the guidelines as supplied by the Department Leaders.
5. At the direction of the Department Leaders, implement restoration procedures and conversion to normal operations at the termination of the emergency.
6. Provide the Department Leaders and other appropriate personnel with updated information on a timely basis, regarding this emergency and its impact.

6.6 Assistant Office and Department Managers

If an emergency is declared by an officer of the company, persons responsible for fulfilling this position shall be responsible for the satisfactory performance of the following assignments:

1. Assist the supervisor by insuring that customer service representatives have secured all cash and negotiable documents, and that they have locked cash drawers and kept the key;
2. Check the immediate area for hazards created by the disaster;
3. Insure that all doors to the company have been locked;
4. Determine if anyone needs medical assistance;
5. Assist anyone in the immediate area who may need guidance in completing their tasks or has other problems.
6. Assist the Office Manager as needed or, in his/her absence, deliver emergency instructions to all staff immediately; assess the risk of potential hazards, or other problems.

6.7 Customer Service Representatives

If an emergency is declared by an officer of the company, persons responsible for fulfilling this position shall be responsible for the satisfactory performance of the following assignments:

1. Immediately secure all cash and negotiable documents, lock cash drawers and keep the key.
2. Check the immediate area for hazards created by the disaster.
3. Determine if anyone needs medical assistance.
4. Report to supervisor immediately for instructions; advise the supervisor of potential hazards, or other pertinent information.
5. Perform tasks as assigned by the supervisor.

7.0 EMERGENCY NOTIFICATION PROCEDURES

7.1 The Disaster Management Team Chairpersons, Coordinators and Department Leaders shall maintain appropriate lists of employees' names, titles, position descriptions, home addresses and home telephone numbers. This information shall be maintained with the Disaster Management Plan for contact, if necessary.

7.2 This personnel information shall be updated at least annually, and shall remain confidential.

7.3 Upon notification of an emergency, the Disaster Management Team Chairpersons shall authorize the notification and immediate response of appropriate personnel. This notification shall be made by phone or in person, and according to one of the three levels of threat as follows:

1. **ALERT 3: STAND-BY**:

   Appropriate personnel are to be notified of the emergency and placed on stand-by status, and shall remain available for emergency Assignment, either by phone or by other means as necessary. If the emergency occurs during business hours:

   1. All personnel are placed on alert.
   2. All appropriate records, equipment and documents are prepared for transfer to secure areas or facilities.
   3. All operations may resume routine functions.

2. **ALERT 2: MINIMUM RESPONSE**:

   Only selected personnel will be called to respond immediately, and all other personnel will remain on stand-by status. If the emergency occurs during business hours:

   1. All personnel are to remain at current locations, until instructed to leave;
   2. All appropriate records, equipment and documents are transferred to secure areas or facilities; and
   3. Business operations involving customers are discontinued, and internal tasks, functions and assignments continue until further instructions are received.

3. **ALERT 1: MAXIMUM RESPONSE**:

   All designated personnel shall respond immediately to their assigned locations, or another location as declared by the Chairpersons. If the emergency occurs during business hours:

   1. The company or any of its facilities may be closed immediately, and all personnel are to remain at their current locations.
   2. All appropriate records, equipment and documents are locked within a secure facility.
   3. All business operations are discontinued, and personnel and customers remain at their locations until escorted from those locations by company officials or emergency services personnel.

7.4 In considering the appropriate response, the Disaster Management Team shall consider which stage(s) of emergency is present:

1. Detection.
2. Reaction.
3. Assessment.
4. Notification.
5. Mobilization.
6. Recovery.
7. Restoration.
8. Resumption.
9. Reconstruction.

7.5 In the event of any activation of this plan, the Disaster Management Team Coordinators shall cause an activity log to be kept at all times during the activation. This log shall be maintained in a chronological sequence of events, and clearly identify the following information:

1. Time activity begun, actions taken, activity terminated.
2. Locations involved.
3. Person maintaining log entries.
4. Person generating information.
5. Person assigned a task, if applicable.
6. Notification of other personnel, if applicable:
    1. Name of person notified.
    2. Time of notification.
    3. Action requested.
    4. Disposition.
7. Unusual events requiring follow-up when normal operations resume; and
8. Termination date/time of emergency activation, including name and title of person making termination.

8.0 EMERGENCY RESPONSE PROCEDURES - ALL PERSONNEL

8.1 Overview

All employees must be able to identify, prioritize and act upon solutions for diminishing the impact of common events that may affect the safety of the company's:

- People
- Employees
- Customers and other persons

Places:

____

____

Facilities;

____

____

Things:

____

Assets; and Records:

____

____

Prior to the onset of an event requiring an emergency response, all employees must become knowledgeable regarding the:

- Persons in charge of emergency response efforts.
- Notification procedures.
- Operation of safety equipment.
- Location and route to established evacuation sites.
- Evacuation procedures.
- Employees' individual emergency response roles.

8.2 Types of Emergency Response Events

8.2.1 Emergency responses are often required when employees encounter one or more of the three (3) types of events likely to result in a disaster, such as:

- Natural or cataclysmic incident caused by internal or external sources:
  - Earthquake or storm
  - Explosion
  - Fire
  - Flood
- Human behavior incident caused by human error or intent:
  - Bomb threat
  - Civil disturbance (riot)
  - Extortion
  - Hostage taking
  - Robbery

- Technological breakdown incident caused by human error or mechanical malfunction:
  - Computer hardware or software failure
  - Gas leak
  - Power outage
  - Telecommunications interruption
  - Toxic waste dispersion
  - Water leak

8.2.2 Determining Factors:

To determine the most appropriate type and level of emergency response:

- Identify the source of the incident.
- Determine if the immediate threat has ended before proceeding.
- Identify and isolate the persons, places and things most likely to be affected by the event.
- Identify the most appropriate options available.
- Choose from among the available options the least dangerous, disruptive or harmful action.
- Act upon your decision.
- If the results are unsatisfactory, choose and act upon another option.

8.2.3 Emergency Response by Event Type

These actions are to be implemented if possible by all personnel, regardless of position description or primary assignment, and only if the employee considers the action safe. The same actions are to be taken if the employee discovers the incident or is assigned to respond to the incident. These actions address the safety of employees and other persons only and do not address security issues regarding facilities, assets or records.

8.3 Disaster Response Guidelines

8.3.1 If a disaster occurs, the discovering employee is to promptly notify his/her immediate supervisor. If the immediate supervisor determines that activation of the plan is required, he/she will promptly:

- Notify Coordinators and Chairpersons, if the disaster is discovered during business hours.
- Notify persons at the selected emergency response phone number, if the disaster is discovered during non-business hours.
- Persons at the selected emergency response phone number maintain emergency phone numbers for Chairpersons, Coordinators and members of the Disaster Management Executive Committee.

- Persons at the selected emergency response phone number will initiate appropriate emergency notifications, and will also initiate an activity log to track events relating to its role in the recovery effort.
- Upon notification by either the reporting supervisor or the Coordinators, the Chairpersons will formally initiate the Disaster Management Plan, and authorize appropriate emergency response to the designated Centers of Operations (Command Centers).
- Chairpersons, Coordinators and appropriate Regional, Office and Department Managers will initiate independent activity logs, to track events and their respective roles during the recovery efforts.
- The Disaster Management Executive Committee will notify available Board members, at a time and in a manner the Chairpersons consider appropriate.
- The Disaster Management Executive Committee will assemble at the designated Command Centers as assigned.
- Coordinators will gather data for the initial damage assessment. Based upon this initial assessment, the Disaster Management Executive Committee will determine the appropriate level and method of response.
- Coordinators will assign emergency contact with phone system representatives to appropriate staff, who will notify company representatives of the disaster.
- Emergency personnel will be assigned for telephone staffing. All personnel assigned to staffing emergency telephones will:
    - Assemble at the designated sites as assigned.
    - Initiate independent telephone activity logs to track events and their respective roles during recovery efforts.
    - Maintain designated emergency lines for contact with Chairpersons, Coordinators, members of the Disaster Management Executive Committee, and appropriate Office and Department Managers.
    - Maintain lines for employees.
    - Lines for routine business and other business functions as available.
- The Disaster Management Executive Committee will notify appropriate Regional, Department and Office Managers to respond promptly to designated Command Centers, or other appropriate sites.
- Emergency personnel will be assigned for on-site telephone staffing. All appropriate Regional, Department and Office Managers and staff personnel assigned to staffing emergency telephones will:
    - Assemble at the designated sites as assigned.
    - Initiate independent telephone activity logs to track events and their respective roles during recovery efforts.

- Maintain designated emergency:
    - Lines for contact with Chairpersons and
    - Coordinators; members of the Disaster Management Executive Committee; and appropriate Office and Department Managers.
    - Lines for employees.
    - Lines for routine business and other business functions as available.

8.3.2 Coordinators will manage responses by personnel, vendors, Regional, Department and Office Managers from the designated Command Centers. Regional, Department and Office Managers will manage on-site responses by the region's emergency services agencies, and other on-site functions as assigned by the Coordinators, including:

- If facilities or assets only have been damaged, representatives from the facilities management, insurance and security functions will be dispatched for on-site damage assessment and control.
- If records have also been damaged or compromised, a legal representative will be dispatched for on- site damage assessment and control.
- If the Data Processing facility has been damaged or is inaccessible, appropriate functions will be shifted to the designated recovery site(s).
- If records have also been damaged or compromised, a legal representative will be dispatched for damage assessment and control.

8.3.3 Based upon the on-site damage assessment, the Disaster Management Executive Committee will determine whether to:

- Close the facility, office or department.
- Temporarily re-assign employees and customer service functions.
- Open the facility, office or department on limited basis, and advise personnel and customers; or
- Open facility, office or department on full-service basis.

8.3.4 Based upon on-site damage assessment and the course of action determined by the Disaster Management Executive Committee, the Chairpersons will initiate an initial press conference and issue a press release. This press release will include statements regarding:

- An initial damage assessment to facilities only.
- Statements regarding injuries, fatalities, and general geographic conditions will be referred to the responsible emergency services agencies.
- The name(s) of the person(s) responsible for managing the recovery effort.
- The names of persons authorized to release information to the media.
- If credit functions are affected, that all customer accounts are safe and protected, and when the credits to those accounts will be available.

- The location designated as a temporary customer-service center, including:
  - The name of the on-site representative(s)
  - The physical address of the location
  - Any "hotline" number designated for customer inquiries
  - The hours of operation

8.3.5 The Disaster Management Executive Committee will initiate the shift from recovery (stabilization) to restoration (of critical functions only), resumption (of all functions), or reconstruction (to the point of disaster). This includes:

- The notification of any appropriate regulatory agencies and insurance carriers of the disaster and a summary of actions taken, if appropriate.
- The issuance of periodic press releases by the Chairpersons, as appropriate.

8.3.6 Based upon the continuing damage assessment, progress reports and recovery efforts, the Disaster Management Executive Committee will determine when to de-activate the disaster status. When the shift to routine operations is initiated:

- Interim policies and procedures will be implemented as appropriate.
- Continuing restoration efforts will be delegated to the on-site representative(s).
- Emergency telephone personnel will be assigned to contact appropriate Regional, Office and Department Managers, and persons at the selected emergency response phone number with instructions de-activation instructions.
- The Disaster Management Executive Committee will conduct its own de-briefing and critique immediately following the plan's de-activation.
- The Chairpersons will issue a press release summary of actions taken and disaster status.
- Notify any appropriate regulatory agencies and insurance carriers of the return to routine operations, or continued restricted activities, if appropriate.
- Advise the Board of Directors of appropriate recovery efforts.
- Appropriate Regional, Office and Department Managers will furnish a written after-action report within 24 hours of the plan's de-activation.
- All independent activity logs will be delivered to Command Center, as attachments to the after-action reports written by Regional, Office and Department Managers.
- Within 24 hours of the plan's de-activation, the Auditor will critique the plan's implementation.
- Meet with the Disaster Management Executive Committee and discuss appropriate findings.

- Make recommendations to revise the Disaster Management Plan, if appropriate.
- Within 24 hours of the Audit review, the Chairpersons will:
  - Notify any appropriate regulatory agencies and insurance carriers of continuing recovery efforts and a summary of actions taken, if appropriate.
  - Prepare and distribute a written summary of the disaster and recovery efforts to any appropriate regulatory agencies, insurance carriers and appropriate personnel, including the Board of Directors.

8.4 Natural or Cataclysmic Incident

8.4.1 Earthquake or storm

- Employees, customers and other persons are to remain in the facility and either take cover under sturdy furniture or stand in an interior doorway.
- If you are outside the facility during the earthquake or storm move away from buildings, trees, overhanging items and power lines.
- Close draperies or blinds over glass windows and doors.
- If it is necessary to move within the building, use stairways and avoid elevators and other mechanical devices.
- Do not use matches, candles or open flames to provide illumination; do not activate any electrical switch: use flashlights only.
- Locate and report the condition of the premises and obvious hazards.
- If you observe structural damage to the facility or an obvious leak immediately turn off all water, gas and electrical power at the main valves and breakers.
- Do not re-activate power switches or valves.
- Do not use the telephone, except to report an emergency.
- Locate and report the condition of all injured persons, and provide first aid assistance as directed.

8.4.2 Explosion

- Activate a fire alarm, or otherwise notify persons within the facility.
- Tell another person to contact the fire department immediately: remain at the site if the area is undamaged.
- If the explosion has damaged the immediate area, abandon the area and follow established evacuation procedures.
- Do not use matches, candles or open flames to provide illumination: use flashlights only.

- If it is necessary to move within the building, use stairways and avoid elevators and other mechanical devices.
- Locate and report the condition of the premises and obvious hazards.
- If you observe structural damage to the facility or an obvious leak immediately turn off all water, gas and electrical power at the main valves and breakers.
- Do not re-activate power switches or valves.
- Do not use the telephone, except to report an emergency.
- Locate and report the condition of all injured persons, and provide first aid assistance as directed.

8.4.3 Fire (materials)

- Activate a fire alarm, or otherwise notify persons within the facility.
- Tell another person to contact the fire department immediately: remain at the site.
- Close all doors and windows in the immediate area.
- Attempt to isolate and control the fire by using an appropriate fire extinguisher or other device.
- If the fire is beyond your immediate control, abandon the area and follow established evacuation procedures.
- Locate and report the condition of the premises and obvious hazards.
- Locate and report the condition of all injured persons, and provide first aid assistance as directed.

8.4.4 Fire (mechanical or device)

1. Activate a fire alarm, or otherwise notify persons within the facility.
2. Tell another person to contact the fire department immediately: remain at the site.
3. Remove combustible materials from the device's immediate area.
4. Turn the electrical power off at the circuit breaker box and do not attempt to unplug the device.
5. Close all doors and windows in the immediate area.
6. Attempt to isolate and control the fire by using an appropriate fire extinguisher or other device.
7. If the fire is beyond your immediate control, abandon the area and follow established evacuation procedures.
8. Locate and report the condition of the premises and obvious hazards.

9 Locate and report the condition of all injured persons, and provide first aid assistance as directed.

10. Do not attempt to re-activate the device until it has been repaired and tested.

8.4.5 Flood

1. Tell another person to contact the fire department immediately: remain at the site.
2. Turn the electrical power off at the circuit breaker box and do not attempt to unplug any device.
3. If flooding is caused by exterior sources attempt to stop the water flow at doors and windows.
4. If flooding is caused by interior sources turn off the main water valve.
5. Locate and report the condition of the premises and obvious hazards.
6. Locate and report the condition of all injured persons, and provide first aid assistance as directed.
7. Do not re-activate any electrical power switch until all power outlets and affected devices have been examined for damage.

8.5 Human Behavior Incident

8.5.1 Bomb threat

1. If you are the person who receives the bomb threat call you are responsible for completing the appropriate form, asking all questions if possible.
2. Your primary purpose is to keep the caller on the telephone line while obtaining as much information as possible.
3. After receiving the initial message, ask the caller to repeat the message for clarity.
4. Tell the caller that you will cooperate with all demands.
5. Ask the caller if a manager may take over for you.
6. Notify a manager of the bomb threat call during or immediately after the call is completed.
7. If you are the person who receives the bomb threat call you will remain the sole contact person for any subsequent calls whenever possible, unless the caller asks to conduct further conversations with a manager.
8. The telephone line on which the bomb threat call was received will remain open for the duration of the event: any notifications will be made on alternate telephone lines by another employee.
9. When possible, two (2) employees will monitor the bomb threat call and each employee is responsible for completing the appropriate form. However, the

employee receiving the bomb threat call is the only person to talk with the caller.

10. You will be removed from the work area and isolated as soon as the bomb threat call is completed. You will be asked to provide all detailed information to the supervisor to whom you are assigned. When this has been completed, you are to resume the responsibility of talking with the caller should he/she call again, if appropriate.
11. Any other employees witnessing unusual or suspicious activities within or around the facility during or after the bomb threat call are to report this information to a manager immediately.
12. No employee will activate a robbery alarm in response to any bomb threat call: a manager will advise the appropriate law enforcement agency of the incident by telephone.
13. A manager will direct a visual search only of the facility for any suspicious or unusual objects, and will determine if it is necessary to abandon the area and follow established evacuation procedures.
14. If a bomb explodes, basic emergency procedures regarding explosions apply.

8.5.2 Civil disturbance (riot)

1. Employees, customers and other persons are to remain in the facility and will be stationed away from doors and windows.
2. Lock all doors and secure all windows.
3. Close draperies or blinds over glass windows and doors and turn out all lights readily visible from outside the facility.
4. Employees and other persons within the facility will not attempt to contact any person causing the disturbance, or otherwise create any action that will draw attention to themselves or the facility.
5. Do not activate a robbery alarm: notify a law enforcement agency by telephone if possible.
6. If it is necessary to move within the building, use stairways and avoid elevators and other mechanical devices.
7. Locate and report the condition of the premises and obvious hazards.
8. Locate and report the condition of all injured persons, and provide first aid assistance as directed.
9. All employees and other persons are to remain within the facility until escorted from the facility by law enforcement personnel, or until they are notified by a manager that they may leave the facility safely.
10. If you are outside and approaching the facility and observe a disturbance, do not enter the facility. Leave the area immediately and report the disturbance to a law enforcement agency.

8.5.3 Extortion

1. If you are the person who receives the extortion threat call you are responsible for completing the appropriate form, asking all questions if possible. Your primary purpose is to keep the caller on the telephone line while obtaining as much information as possible.
2. After receiving the initial message, ask the caller to repeat the message for clarity.
3. Tell the caller that you will cooperate with all demands.
4. Ask the caller if a manager may take over for you.
5. Notify a manager of the extortion threat call during or immediately after the call is completed.
6. If you are the person who receives the extortion threat call you will remain the sole contact person for any subsequent calls whenever possible, unless the caller asks to conduct further conversations with a manager.
7. The telephone line on which the extortion threat call was received will remain open for the duration of the event: any notifications will be made on alternate telephone lines by another employee.
8. When possible, two (2) employees will monitor the extortion threat call and each employee is responsible for completing the appropriate form. However, the employee receiving the extortion threat call is the only person to talk with the caller.
9. You will be removed from the work area and isolated as soon as the extortion threat call is completed. You will be asked to provide all detailed information to the supervisor to whom you are assigned. When this has been completed, you are to resume the responsibility of talking with the caller should he/she call again, if appropriate.
10. Any other employees witnessing unusual or suspicious activities within or around the facility during or after the extortion threat call are to report this information to a manager immediately.
11. No employee will activate a robbery alarm in response to any extortion threat call: a manager will advise the appropriate law enforcement agency of the incident by telephone.

8.5.4 Hostage taking

If a hostage is taken inside the facility during a robbery or is brought into the facility during an extortion attempt, basic emergency procedures regarding robberies apply, except or in addition to:

1. Do not activate a robbery alarm: notify a law enforcement agency by telephone if possible.

2. If you are the employee contacted by the hostage taker, encourage the hostage-taker to leave the hostage in the facility after all demands are met.
3. Do not offer to exchange yourself for or accompany the hostage.

8.5.6 Robbery

1. If you are the person being robbed your primary purpose is to help the robber complete the robbery successfully.
2. Do not attempt to involve or aid another employee or other persons during the robbery.
3. Keep your hands in plain view, and avoid rapid movements or actions that may be misinterpreted.
4. Tell the robber that you will cooperate with all demands, and repeat the demands back to the robber before acting, if possible.
5. Give the robber no more and no less than he/she demands.
6. Include the "bait" money and dye pack with other currency. If the robber tells you not to include those items, comply with the robber's demands.
7. If the robber tells you go to other areas within the facility and continue to help him/her, comply with the robber's demands.
8. If the robber physically assaults you during the robbery, you must decide whether to allow this assault to continue: an employee never loses the basic right to self-defense.
9. If the robber demands that you accompany him/her outside the facility, you must decide whether to comply: an employee never loses the basic right to self-defense. You will be at greater personal risk if you leave the facility with the robber.
10. Activate the surveillance cameras and alarms only when it is safe to do so.
11. Keep any demand notes received, and remember which areas of the facility were touched by the suspect.
12. You will be removed from the work area and isolated as soon as the robbery is completed. You will be asked to provide all detailed information to the supervisor to whom you are assigned.
13. A manager will lock the facility's doors and protect the crime scene as soon as it is safe to do so.
14. Locate and report the condition of the premises and obvious hazards.
15. Locate and report the condition of all injured persons, and provide first aid assistance as directed.
16. Any other employees witnessing unusual or suspicious activities within or around the facility during or after the extortion threat call are to report this information to a manager immediately.

17. If you are an observer rather than a participant in the robbery, do not interfere with the participants.
18. Be observant concerning the robber's physical characteristics and observe the suspect's direction of travel, number of accomplices and descriptions and any other pertinent information.
19. Activate the surveillance cameras and alarms or contact the law enforcement agency by telephone only when it is safe to do so.

8.6 Technological Breakdown Incident

8.6.1 Computer hardware or software failure

- Tell another person to contact the data processing manager - remain at the site.
- If the failure involves computer hardware, immediately remove combustible materials from the device's immediate area. Then follow the same procedures for either hardware or software failure.
- Do not turn the device off unless the failure will create additional damage to hardware or software if the device is left operating. This will help the data processing manager trace the source of the failure.
- Turn the device off if the failure will create additional damage to hardware or software if the device is left operating. Decide whether to turn off electrical power at the circuit breaker box or by using the device's power switch.
- Do not attempt to re-activate the device until it has been repaired and tested.

8.6.2 Gas leak

- Tell another person to contact the fire department immediately and to notify persons within the facility.
- Do not activate a fire alarm or any electrical device.
- Do not use matches, candles or open flames to provide illumination: use flashlights only.
- Close all doors and windows in the immediate area, and abandon the area and follow established evacuation procedures.
- If it is necessary to move within the building, use stairways and avoid elevators and other mechanical devices.
- Turn the electrical power off at the circuit breaker box and the gas supply at the gas valve.
- Do not use the telephone at the affected site: use another telephone only to report an emergency.
- Locate and report the condition of the premises and obvious hazards.

- Locate and report the condition of all injured persons, and provide first aid assistance as directed.
- Do not re-activate power switches or gas valves until the source of the gas leak has been identified and repaired.

8.6.3 Power outage

- Tell another person to notify persons within the facility. If necessary, also contact the fire department.
- Do not activate a fire alarm or any electrical device.
- Do not use matches, candles or open flames to provide illumination: use flashlights only.
- Do not use the telephone except to report an emergency.
- If it is necessary to move within the building, use stairways and avoid elevators and other mechanical devices.
- Turn off all surge protectors and other electrical devices to prevent damage from an electrical surge when power is restored.
- Contact the power company and determine the cause of the failure, and the estimated time of return of service.
- If a power failure results in the accompanying failure of the company's alarms, close the facility until power is restored and all systems are returned to normal functions.
- Determine if employees or other persons are trapped within rooms with electronic access, or within elevators; if such persons are located, immediately notify the emergency response agency of the number of persons and their location(s).
- Locate and report the condition of the premises and obvious hazards.
- Locate and report the condition of all injured persons, and provide first aid assistance as directed.
- Do not re-activate power switches or electrical devices until the source of the power outage has been identified and repaired.
- Use an emergency generator to supply electric power only after determining that the source of the outage was caused by the power company, and that the facility's electrical system is safe to operate using a generator.

8.6.4 Telecommunications interruption

- If an interruption to basic telecommunications services occurs, consider applying appropriate basic emergency procedures regarding a power outage until you determine the source of the interruption.

- Tell another person to notify persons within the facility: remain at the site of the interruption.
- Do not activate a fire or robbery alarm.
- If you must use the telephone to report an emergency, use a nearby public telephone if one is operable.
- Contact the telecommunications company and determine the cause of the failure, and the estimated time of return of service.
- If a telecommunications failure results in the accompanying failure of the company's alarms, close the facility until power is restored and all systems are returned to normal functions.

8.6.5 Toxic waste dispersion

- Tell another person to notify persons within the facility and to contact the fire department immediately.
- If the toxic hazard is outside the facility:
  - Employees, customers and other persons are to remain in the facility and will be stationed away from doors and windows.
  - Turn all electrical power off at the circuit breaker box.
  - Lock all doors and secure all windows.
  - Open draperies or blinds over glass windows and doors so that employees and other persons are readily visible from outside the facility.
  - Do not activate a robbery alarm: notify a law enforcement agency by telephone if possible.
- If it is necessary to move within the building, use stairways and avoid elevators and other mechanical devices.
- Locate and report the condition of the premises and obvious hazards.
- Locate and report the condition of all injured persons, and provide first aid assistance as directed.
- All employees and other persons are to remain within the facility until escorted from the facility by law enforcement or fire personnel, or until they are notified by a manager that they may leave the facility safely.

8.6.6 Toxic hazard inside the facility

- Turn all electrical power off at the circuit breaker box.
- If the toxic hazard has affected the immediate area, abandon the area and follow established evacuation procedures, locking all doors and securing all windows upon departure.

- If it is necessary to move within the building, use stairways and avoid elevators and other mechanical devices.
- Do not activate a robbery alarm: notify a law enforcement agency using an available public telephone.
- Locate and report the condition of the premises and obvious hazards.
- Locate and report the condition of all injured persons, and provide first aid assistance as directed.
- Do not re-activate any power switch at the circuit breaker box until all sources of ventilation and appropriate surfaces have been examined for continuing toxic exposure.

8.6.7 Water leak

- Tell another person to contact the fire department immediately: remain at the site.
- Turn the electrical power off at the circuit breaker box and do not attempt to unplug any device.
- If the water leak is caused by interior sources turn off the main water valve.
- Locate and report the condition of the premises and obvious hazards.
- Locate and report the condition of all injured persons, and provide first aid assistance as directed.
- Do not re-activate any electrical power switch until all power outlets and affected devices have been examined for damage.

9.0 SERVICE AGREEMENTS, EMERGENCY SERVICES AGENCIES AND COMMUNITY RESOURCES

9.1 The Disaster Management Team Coordinators shall continually investigate appropriate vendors and community resources to provide emergency services to the company. The company shall rely, whenever possible, upon established vendors, internal devices, personnel and procedures to administer the recovery effort.

9.2 The Disaster Management Team Coordinators shall continue to seek appropriate service agreements with other companies, associations and service companies, to provide backup services until the company has resumed normal operations. As those agreements are developed, they shall be added to the Disaster Management Plan, updated and verified at least annually.

9.3 The Disaster Management Team Coordinators shall also be responsible for developing and maintaining a comprehensive listing of all emergency services and related agencies within each county, or other appropriate geographic region, serviced by the company.

Wherever possible, two (2) similar agencies, resources and vendors will be listed. This listing shall be updated and verified at least annually.

9.4 This listing shall focus upon those services and resources likely to assist with recovery efforts regarding:

1. Personnel: Notification, basic needs, transportation, re-allocation, replacement, management, media relations, temporary lodging.
2. Customers: Notification, marketing, public relations, cash-handling, credit approval.
3. Telecommunications: Internal, external, public access, facsimile machine and modem, cellular.
4. Emergency services: Law enforcement, fire, medical, non-emergency governmental agencies.
5. Facilities: Power, utility agencies, repair and Replacement contractors.
6. Data Processing: Item processing, capture, proof, ACH, customer and vendor information, critical records, office PCs and peripherals.
7. Equipment and supplies: Repair, replacement, special needs, delivery.

■ EMERGENCY SERVICES AND AGENCIES ■

EMERGENCY SERVICES

Name: Police Department

City:

County:

Emergency Phone: 911

Business Phone:

Name: Sheriff's Department

County:

Emergency Phone: 911

Business Phone:

Name: Fire Department

City:

County:

Emergency Phone: 911

Business Phone:

Name: Paramedic/Rescue #1

City:

County:

Emergency Phone: 911

Business Phone:

Name: Air Ambulance

County:

Emergency Phone:

Business Phone:

Name: Private Ambulance #1

County:

Emergency Phone:

Business Phone:

Name: Private Ambulance #2

County:

Emergency Phone:

Business Phone:

HOSPITAL / URGENT CARE FACILITY

Name: Hospital #1

Address:

City:

County:

Emergency Phone:

Business Phone:

Name: Hospital #2

Address:

City:

County:

Emergency Phone:

Business Phone:

Name: Hospital #3

Address:

City:

County:

Emergency Phone:

Business Phone:

Name: Urgent Care Facility #1

Address:

City:

County:

Emergency Phone:

Business Phone:

Name: Urgent Care Facility #2

Address:

City:

County:

Emergency Phone:

Business Phone:

EMERGENCY STAGING FACILITIES AND SHELTERS

Name: American Red Cross

Address:

City:

County:

Emergency Phone:

Business Phone:

Staging location:

Name: Community Center #1

Address:

City:

County:

Emergency Phone:

Business Phone:

Staging location:

Name: Community Center #2

Address:

City:

County:

Emergency Phone:

Business Phone:

Staging location:

Name: National Guard Center

Address:

City:

County:

Emergency Phone:

Business Phone:

Staging location:

Name: Veterans Memorial Building

Address:

City:

County:

Emergency Phone:

Business Phone:

Staging location:

DISASTER INFORMATION

Name: Medical Emergency Information Hotline

County:

Emergency Phone:

Name: Office of Emergency Services (City)

Address:

City:

County:

Emergency Phone:

Business Phone:

Staging location:

Name: Office of Emergency Services (County)

Address:

City:

County:

Emergency Phone:

Business Phone:

Staging location:

Name: Office of Emergency Services (State)

Address:

City:

County:

Emergency Phone:

Staging location:

COMMUNICATIONS

Name: Company for remote contact #1

Address:

Phone 1:

Phone 2:

FAX:

Name: Company for remote contact #2

Address:

Phone 1:

Phone 2:

FAX:

Name: Mobile telephone #1

Address:

Phone 1:

Phone 2:

Cellular access numbers:

Name: Mobile telephone #2

Address:

Phone 1:

Phone 2:

Cellular access numbers:

Name: Newspaper #1

City:

County:

Emergency Phone:

Business Phone:

FAX:

Name: Newspaper #2

City:

County:

Emergency Phone:

Business Phone:

FAX:

Name: Public telephone #1 (3 for each facility)

Address or location:

Phone:

Name: Public telephone #2 (3 for each facility)

Address or location:

Phone:

Name: Public telephone #3 (3 for each facility)

Address or location:

Phone:

Name: Radio Station #1

City:

County:

Emergency Phone:

Business Phone:

FAX:

Name: Radio Station #2

City:

County:

Emergency Phone:

Business Phone:

FAX:

Name: Television Station #1

City:

County:

Emergency Phone:

Business Phone:

FAX:

Name: Television Station #2

City:

County:

Emergency Phone:

Business Phone:

FAX:

SECURITY

Name: Alarm Company #1

City:

County:

Emergency Phone:

Business Phone:

Name: Alarm Company #2

City:

County:

Emergency Phone:

Business Phone:

Name: Guard Company #1

City:

County:

Emergency Phone:

Business Phone:

Name: Guard Company #2

City:

County:

Emergency Phone:

Business Phone:

TRANSPORTATION

Name: Airport/Municipal (City)

Address:

City:

County:

Emergency Phone:

Business Phone:

Name: Airport/Regional (County)

Address:

City:

County:

Emergency Phone:

Business Phone:

Name: Bus (City)

City:

County:

Emergency Phone:

Business Phone:

Name: Bus (County)

City:

County:

Emergency Phone:

Business Phone:

Name: Cab/Taxi #1

City:

County:

Emergency Phone:

Business Phone:

Name: Cab/Taxi #2

City:

County:

Emergency Phone:

Business Phone:

Name: Car Rental Agency #1

City:

County:

Emergency Phone:

Business Phone:

Name: Car Rental Agency #1

City:

County:

Emergency Phone:

Business Phone:

Name: Cash - Records Transport/Supplemental #1

City:

County:

Emergency Phone:

Business Phone:

Name: Cash - Records Transport/Supplemental #2

City:

County:

Emergency Phone:

Business Phone:

Name: Moving Company #1

City:

County:

Emergency Phone:

Business Phone:

Name: Moving Company #2

City:

County:

Emergency Phone:

Business Phone:

CITY / COUNTY OFFICES

Name: Administrative Offices (City)

City:

County:

Emergency Phone:

Business Phone:

Name: Administrative Offices (County)

City:

County:

Emergency Phone:

Business Phone:

Name: Air Quality Control Offices (County)

City:

County:

Emergency Phone:

Business Phone:

Name: Animal Control (City)

City:

County:

Emergency Phone:

Business Phone:

Name: Animal Control (County)

City:

County:

Emergency Phone:

Business Phone:

Name: Building Inspector (City)

City:

County:

Emergency Phone:

Business Phone:

Name: Building Inspector (County)

City:

County:

Emergency Phone:

Business Phone:

Name: Hazardous Materials Team (County)

City:

County:

Emergency Phone:

Business Phone:

Name: Health (City)

County:

Emergency Phone:

Business Phone:

Name: Health (County)

City:

County:

Emergency Phone:

Business Phone:

Name: Health and Safety Inspector (County)

City:

County:

Emergency Phone:

Business Phone:

Name: Mental Health (City)

City:

County:

Emergency Phone:

Business Phone:

Name: Mental Health (County)

City:

County:

Emergency Phone:

Business Phone:

Name: Parks and Recreation (City)

County:

Emergency Phone:

Business Phone:

Name: Parks and Recreation (County)

County:

Emergency Phone:

Business Phone:

Name: Public Information Office (City)

City:

County:

Emergency Phone:

Business Phone:

Name: Public Information Office (County)

City:

County:

Emergency Phone:

Business Phone:

Name: Public Works (City)

City:

County:

Emergency Phone:

Business Phone:

Name: Public Works (County)

City:

County:

Emergency Phone:

Business Phone:

Name: Schools Information (City)

City:

County:

Emergency Phone:

Business Phone:

Name: Schools Information (County)

City:

County:

Emergency Phone:

Business Phone:

Name: Transportation/Roads (City)

City:

County:

Emergency Phone:

Business Phone:

Name: Transportation/Roads (County)

City:

County:

Emergency Phone:

Business Phone:

FEDERAL OFFICES

Name: Department of Corporations

Address:

Emergency Phone:

Business Phone:

Name: Federal Bureau of Investigation

Address:

Emergency Phone:

Business Phone:

UTILITIES

Name: Electric

City:

County:

Emergency Phone:

Business Phone:

Name: Gas

City:

County:

Emergency Phone:

Business Phone:

Name: Sanitation/Sewer

City:

County:

Emergency Phone:

Business Phone:

Name: Telephone

City:

County:

Emergency Phone:

Business Phone:

Name: Water

City:

County:

Emergency Phone:

Business Phone:

CONTRACTORS / VENDORS

Name: Building/Reconstruction Contractor

City:

County:

Emergency Phone:

Business Phone:

Name: Electrical Contractor

City:

County:

Emergency Phone:

Business Phone:

Name: Glass Contractor

City:

County:

Emergency Phone:

Business Phone:

Name: Grounds-keeping Contractor

City:

County:

Emergency Phone:

Business Phone:

Name: Hazardous Materials Response Contractor

City:

County:

Emergency Phone:

Business Phone:

Name: Health and Safety Contractor

City:

County:

Emergency Phone:

Business Phone:

Name: Janitorial Contractor

City:

County:

Emergency Phone:

Business Phone:

Name: Locksmith and Vault Contractor

City:

County:

Emergency Phone:

Business Phone:

Name: Office Supplies and Forms Contractor

City:

County:

Emergency Phone:

Business Phone:

Name: Plumbing Contractor

City:

County:

Emergency Phone:

Business Phone:

Name: Roofing Contractor

City:

County:

Emergency Phone:

Business Phone:

Name: Utility Contractor

City:

County:

Emergency Phone:

Business Phone:

EQUIPMENT

Name: Applications

City:

County:

Emergency Phone:

Business Phone:

Name: Mainframe computers

City:

County:

Emergency Phone:

Business Phone:

Name: Office equipment other than PCs

City:

County:

Emergency Phone:

Business Phone:

Name: Personal computers

City:

County:

Emergency Phone:

Business Phone:

Name: Software

City:

County:

Emergency Phone:

Business Phone:

10.0 CENTERS OF OPERATIONS

10.1 The designated Centers of Operations are to be used as command centers, in the event of an emergency activation of this plan. These requirements have been considered:

1. Appropriateness of the facility to house specific operations.
   Convenience to customers, if this is a customer-service facility; and Building security and vault availability if this is a temporary cash-handling facility.
2. Adequacy of public transportation, and availability of sufficient roadways, highways and surface streets.
3. Adequacy of land-line and cellular phone service.
4. Availability of adjacent temporary housing and food service facilities.

10.2 The company's primary designated Center of Operations is located at:

The **Main Office** - <Building Name>

Address: ______________________________

City: ______________________________

State: __________ ZIP: ______________

Phone: ______________ FAX: ______________

10.3 The company's secondary Centers of Operations are, in designated order:

1. **Secondary Center #1**

   Address: ______________________________

   City: ______________________________

   State: __________ ZIP: ______________

   Phone: ______________ FAX: ______________

2. **Secondary Center #2**

   Address: ______________________________

   City: ______________________________

   State: __________ ZIP: ______________

   Phone: ______________ FAX: ______________

3. **Secondary Center #3**

   Address: ______________________________

   City: ______________________________

   State: __________ ZIP: ______________

   Phone: ______________ FAX: ______________

10.4 The following supplies and equipment shall be stored at the primary and secondary Centers of Operations, and at each office, department and facility as appropriate:

1. Current full copy of the company's Disaster Management Plan and operations procedures for each office, facility and function.
2. Building plans for each facility, including primary control points for all utilities, if applicable.
3. Complete and current list of all personnel, including names, addresses, and phone numbers. Current position descriptions indicating routine duties and responsibilities shall be included.
4. Complete and current list of all equipment and serial numbers; replacement information; names and phone numbers of designated vendors; and service agreement information.
5. Complete and current list of all emergency services, community resources and related agencies within each county, or other geographic region, serviced by the company.
6. Food, water and medical supplies as appropriate.
7. Battery-powered radios capable of receiving emergency instructions; fire extinguishers; flashlights; and basic tools as appropriate.
8. Master keys, alarm access codes and other security information for each office, facility and the other Centers of Operations.
9. Stationery items, numbered forms and negotiable documents as appropriate.
10. Assorted writing materials, instruments and business tools as appropriate.
11. Non-electrical mechanical devices as appropriate.
12. Chronological log forms to document all actions taken during the emergency.
13. Any other items considered appropriate during the initial and annual review process.

10.5 In addition to the primary and secondary Centers of Operations, a storage site may be maintained at a location distant from those Centers. This storage site may contain items acquired in addition or in place of the items stored at the Centers of Operations. These items are to be determined by the Disaster Management Team Coordinators. This storage site is:

Building name: ____

Address: ____

City: ____

State: ____ ZIP: ______________

Phone: __________________________ FAX: ___________________

Space #: ______

Keys are maintained by: ____________________________________

11.0 TESTING AND TRAINING REQUIREMENTS

11.1 All procedures developed and maintained for this plan shall be subjected to a continuous, controlled testing procedure, monitored by the Disaster Management Team Coordinators.

11.2 On completion of each testing procedure, the Disaster Management Team Coordinators shall furnish the Disaster Management Team Chairpersons with a written evaluation of the results of each test.

11.3 Based upon periodic written evaluations, the necessary procedures will either be re-developed and tested, or be accepted as written.

11.4 Upon the adoption of any procedures in this plan, appropriate personnel will be trained regarding the methods of properly implementing those procedures.

11.5 This training program shall be developed, delivered and maintained by the Department Leaders, with the assistance of the Disaster Management Team Coordinators.

11.6 Completion of the Disaster Management Plan training program shall be mandatory for all employees, regardless of job title or position, and may include the Board of Directors. Newly-hired employees shall be trained at the time of initial orientation, and current employees shall be trained on at least an annual basis.

11.7 The Department Leaders shall document all policies and procedures affected by this program, in writing. This documentation shall include, but is not limited to:

1. Course objectives.
2. Course outline.
3. Lesson plan.
4. Handout materials.
5. Audio-visual aids, if appropriate.
6. Training schedules, including dates, times, material addressed, and names/titles of persons attending.

11.8 In the continual design, implementation and delivery of this training program, the Department Leaders shall use the following criteria:

1. Instruction of shall accurately reflect the company's policies and procedures.
2. Instruction regarding the appropriate methods of operating safety and related equipment shall accurately reflect the manufacturers' specifications.

Notes

12.0 ANNUAL RISK ASSESSMENT AND EVALUATION

12.1 All potential risks to the company shall be evaluated on at least an annual basis. The Annual Assessment shall be conducted by each Department Leader, and the written results and recommendations shall be forwarded to the Disaster Management Team Coordinators.

12.2 On receipt of these written results and recommendations, the Disaster Management Team Coordinators shall also issue a report of findings to the Disaster Management Team Chairpersons. This report shall include:

1. An overview of the status of each department regarding compliance with policies and procedures;
2. A projection regarding any new equipment, personnel, policies or procedures required for the next year, to meet the requirements of this plan; and
3. Any necessary change in current policies or procedures not addressed elsewhere, and an accompanying rationale or explanation.

12.3 In conducting this annual assessment, the following criteria shall be used by both Department Leaders and the Disaster Management Team Coordinators:

1. Any potential risk shall be identified, whether already addressed or newly anticipated. If the identified risk has already been successfully addressed, no further assessment or recommendations shall be required. Any newly identified risk shall be fully addressed.
2. Identified risks may be categorized as natural or technological disasters, or caused by human actions. The listed occurrences have been identified as potentially creating an adverse impact upon the company's operations:
    1. Fire
    2. Power outage
    3. Water leak
    4. Gas leak
    5. Bomb threat
    6. Earthquake
    7. Flood
    8. Toxic (hazardous) waste dispersion
    9. Civil disturbance
    10. Bomb explosion
    11. Sabotage
    12. Tornado
    13. Hurricane
    14. Nuclear attack

12.4 Each of these potential risks shall be evaluated according to the following guidelines:

1. What is your ability to predict this occurrence?
2. What is the likely frequency of this occurrence?
3. If applicable, what is the estimated elapsed time between your prediction and the onset of this occurrence?
4. If applicable, what is the estimated elapsed time between your discovery of the occurrence and any injury or damage?
5. What is the possibility of your controlling or diminishing the impact of this occurrence?
6. What is the anticipated duration of this occurrence?
7. What is the projected impact to the company if this event occurs, regarding:
    1. Personnel injury or working conditions.
    2. Customer confidence and the company's image.
    3. Security measures necessary to insure adequate protection of personnel, customers, and facilities, assets and records.
    4. Legal issues surrounding the company's inability to fulfill agreements and obligations, process documents and other information.
    5. Requirements to backup or substitute data or support services.
    6. The effect upon related functions.
    7. Recovery time.
    8. Money lost.
    9. Equipment lost.

12.5 Once potential risks have been identified and prioritized, the Disaster Management Team shall:

1. Determine potential short and long-range solutions.
2. Decide upon the most appropriate primary and secondary solutions.
3. Test and evaluate these solutions for effectiveness.
4. Include these solutions in the Disaster Management Plan, and related testing and training programs.

12.6 All original records of the processes used to compile the information resulting in these policies and procedures shall be retained by the Disaster Management Team Coordinators, in a location separate from the Disaster Management Plan. These records shall include the names and positions of all appropriate persons involved in the continued planning and writing of this plan.

12.7 The following personnel are charged with the responsibility of the conception, development, initiation and maintenance of the Disaster Management Plan, as this plan pertains to their departments and/or responsibilities:

DEPARTMENT NAME: Administration
MANAGER:
DEPARTMENT NAME: Data Processing
MANAGER:
DEPARTMENT NAME: Loan Services
MANAGER:
DEPARTMENT NAME: Operations
MANAGER:
DEPARTMENT NAME:
MANAGER:
DEPARTMENT NAME:
MANAGER:
DEPARTMENT NAME:
MANAGER:
DEPARTMENT NAME:
MANAGER:
DEPARTMENT NAME:
MANAGER:
DEPARTMENT NAME:
MANAGER:

12.8 All insurance policies, and service and vendor agreements shall be reviewed at least annually. If policies and agreements are satisfactory, existing agreements shall be renewed. New vendors will be investigated to replace any unsatisfactory agreements. The listing of these agreements, emergency services agencies, and community resources shall be reviewed and updated at least annually, and added to Section 9.0 of the Disaster Recovery plan.

12.9 The Disaster Management Team Coordinators shall also be responsible for researching and developing a listing of other product and service companies and individuals who may be utilized on an interim basis. This listing shall be developed within each county, or other appropriate geographic area, serviced by the company, to provide backup or restoration services until the company has resumed normal operations. This listing shall be reviewed and updated at least annually, and added to Section 9.0 of the Disaster Recovery plan.

13.0 DEPARTMENT RECOVERY PLAN - Administration

Department: **Administration**

Address: ______________________________

City: ______________________________

State: ________ ZIP: ______________

Phone: ______________________________

FAX: ______________________________

Department Leader: ______________________________

Title: ______________________________

Assistant Department Leader: ______________________________

Title: ______________________________

13.1 Maintaining contact with members of the company's Disaster Management Team during a disaster is critical to a successful Department recovery effort. Usual business phone numbers are listed below, and these numbers should be used for all primary contact with Team members.

Disaster Management Chairpersons:

Name: ______________________________ Phone: __________

Emergency Phone: ______________________________

Cellular Phone: ______________________________

Name: ______________________________ Phone: __________

Emergency Phone: ______________________________

Cellular Phone: ______________________________

Disaster Management Coordinators:

Name: ______________________________ Phone: __________

Emergency Phone: ______________________________

Cellular Phone: ______________________________

Facilities/Equipment/Supplies / Transportation/Telecommunications:

Name: ______________________________ Phone: ______________

Emergency Phone: ______________________________

Cellular Phone: ______________________________

Other Department Leaders:

Name: ______________________________ Phone: ______________

Emergency Phone: ______________________

Cellular Phone: ________________________

Name: ______________________________ Phone: ______________

Emergency Phone: ______________________

Cellular Phone: ________________________

Name: ______________________________ Phone: ______________

Emergency Phone: ______________________

Cellular Phone: ________________________

Name: ______________________________ Phone: ______________

Emergency Phone: ______________________

Cellular Phone: ________________________

Name: ______________________________ Phone: ______________

Emergency Phone: ______________________

Cellular Phone: ________________________

13.2 In the event that normal phone lines are not functional, alternate communications may be available by using public phones (Bell system only). The public phones most readily accessible by Department personnel are:

Public Phone # 1: ______________________________

Location: ______________________________

Public Phone # 2: ______________________________

Location: ______________________________

Public Phone # 3: ______________________________

Location: ______________________________

13.3 The company has established a hotline phone number for emergency use by all employees. Use of this special number is restricted to disaster recovery efforts and emergency notifications only, and is not to be used for any other purpose.

Disaster Recovery Hotline number for employees only: ______________

Security Alarm Company: ______________________________

13.4 The Department Leader or designate is to immediately take the following actions if a disaster occurs:

1. Assess any injury or damage to employees, customers, and facilities.
2. Temporarily close and secure the facility, if necessary.
3. Contact appropriate emergency services, if necessary.
4. Begin documenting the effects of the disaster and actions taken; secure all assets and records.
5. First attempt to contact the company's Disaster Management Team Coordinators or Chairpersons with a damage assessment and actions taken report, and act upon instructions received.
6. If all documented attempts to communicate with Coordinators and Chairpersons have failed, the Department Leader or designate is authorized to initiate reasonable and prudent responses necessary to minimize potential:
    1. Injuries to personnel and customers.
    2. Damage to facilities.
    3. Loss of assets and records.

13.5 The critical functions of the Department are to be serviced before the performance of any other task. The critical Department functions are:

1. Administrative operations.
2. Business operations.
3. Board of Directors.
4. Media relations.
5. Security.
6. Stock/shareholder relations.

13.6 The accessory functions of the Department are to be addressed only after all critical functions have been performed. The accessory functions of the Department are:

1. Credit Administration.
2. Personnel.
3. Accounting / Auditing.
4. Maintain custody of important company records.
5. Insurance policies and Board-related matters.

13.7 Description of Department Leader's duties and responsibilities during a disaster:

1. Ultimately responsible for overall Department operations, including all personnel, customers, facilities and assets.
2. Department Security Officer, Department Compliance Officer and Department Disaster Management Team Coordinator; interior and exterior Department physical security and appearance.
3. Ensure adequate supervision for all personnel and functions while absent from the company or unavailable for contact; operational quality control.
4. Respond to and comply with all regulations, policies and procedures regarding Department operations; prepare reports as required.
5. Customer service and public relations; resolving customer complaints; approve unusual or unique transaction when no other person has that immediate authority; provide information to supervisor for media relations and all requests for interviews from the press, radio and television.
6. Provide appropriate members of the Disaster Management Team with accurate and timely information updates regarding the Department's recovery efforts.
7. Other duties and responsibilities as required.

13.8 Description of Assistant Department Leader's duties and responsibilities:

1. Perform all duties and responsibilities of the Department Leader in his/her absence or because of his/her unavailability.
2. Ensure dual custody requirements are maintained for all functions; maintain key, employee information log for dual custody assignments; enforce employee and functional security procedures, Department opening and closing procedures.
3. Manage day-to-day operational functions, and directly supervise all staff personnel.
4. Other duties and responsibilities as required.

13.9 If a disaster occurs during working hours, the staff will evacuate the facility and gather at an off-site location.

PRIMARY: ______________________________

Location: ______________________________

SECONDARY: ______________________________

Location: ______________________________

A diagram of the facility and designated emergency staging areas should be placed at the end of this section.

13.10 If the Department is not able to be used, Department operations will immediately shift to these alternate sites:

PRIMARY: ____________________

Location: ____________________

Address: ____________________

City: ____________________

State: ____________ ZIP: ____________

Phone: ____________________

FAX: ____________________

SECONDARY: ____________________

Location: ____________________

Address: ____________________

City: ____________________

State: ____________ ZIP: ____________

Phone: ____________________

FAX: ____________________

13.11 If the Department is still operable, this checklist describes the functions or sections upon which you will concentrate recovery efforts, and in what order. Before opening the Department:

1. Assess safety considerations for employees and customers.
2. Coordinate with emergency services agencies, if necessary.
3. Conduct a damage assessment of the building; determine levels of operation or restoration time for electricity, telephones, water, computer system.
4. Ensure all areas of responsibility are staffed.
5. Ensure equipment and supplies are available.
6. Arrange for the safe relocation of all records and equipment, if necessary.

13.12 If it is safe to open the Department, re-establish:

1. Employee, customer, facility, assets and records security.
2. Contact with the Board of Directors.
3. Contact with media representatives.
4. Corporate files and financial records.
5. Personnel and fixed asset records.

6. Accounting records.
7. Insurance policies and documentation.

13.13 The Department requires these logistical factors to be available to perform critical functions:

Square feet: __________

Maximum number of employees: __________

Maximum number of customers: __________

Special relocation needs in the event the facility is unable to support Department operations are: ______________________________

____

13.14 A listing of all emergency services, personnel and equipment available to this Department is located in Section 9.0.

13.15 Additional guidelines to assist disaster recovery efforts for this Department are:

Written operations procedures: ______________________________

Location stored: ______________________________

Container description: ______________________________

13.16 All service agreements and vendor information are available by contacting:

____

____

____

13.17 Additional office supplies, emergency equipment and survival supplies to assist disaster recovery efforts for this Department are:

____

____

____

Emergency medical supplies available:

____

____

____

Location stored:

____

____

Container description:

____

____

13.18 An appropriate supply of the following forms is to be maintained:

____

____

____

____

____

____

____

____

PERSON COMPLETING: ____________________________

TITLE: ______________________________________

DATE: ________________ FORWARDED TO: ____________________

Notes

14.0 DEPARTMENT RECOVERY PLAN – Cash handling facility

Office: **Cash handling facility**

Address:

City:

State ZIP:

Phone:

FAX:

Office Leader:

Title: Office Manager

Assistant Office Leader:

Title:

14.1 Maintaining contact with members of the company's Disaster Management Team during a disaster is critical to a successful Office recovery effort. Usual business phone numbers are listed below, and these numbers should be used for all primary contact with Team members.

Disaster Management Chairpersons:

Name: Phone:

Emergency Phone:

Cellular Phone:

Name: Phone:

Emergency Phone:

Cellular Phone:

Disaster Management Coordinators:

Name: Phone:

Emergency Phone:

Cellular Phone:

Name: Phone:

Emergency Phone:

Cellular Phone:

Facilities/Equipment/Supplies/Transportation/Telecommunications:

Name: Phone:

Emergency Phone:

Cellular Phone:

Other Department Leaders:

Name: Phone:

Emergency Phone:

Cellular Phone:

Name: Phone:

Emergency Phone:

Cellular Phone:

Name: Phone:

Emergency Phone:

Cellular Phone:

Name: Phone:

Emergency Phone:

Cellular Phone:

Name: Phone:

Emergency Phone:

Cellular Phone:

Name: Phone:

Emergency Phone:

Cellular Phone:

14.2 In the event that normal phone lines are not functional, alternate communications may be available by using public phones (Bell system only). The public phones most readily accessible by Office personnel are:

Public Phone # 1:

Location:

Public Phone # 2:

Location:

Public Phone # 3:

Location:

14.3 The company has established a hotline phone number for emergency use by all employees. Use of this special number is restricted to disaster recovery efforts and emergency notifications only, and is not to be used for any other purpose.

Disaster Recovery Hotline number for employees only:

Security Alarm Company:

14.4 The Office Leader or designate is to immediately take the following actions if a disaster occurs:

1. Assess any injury or damage to employees, customers, and facilities.
2. Temporarily close and secure the facility, if necessary.
3. Contact appropriate emergency services, if necessary.

4. Begin documenting the effects of the disaster and actions taken; secure all assets and records.
5. First attempt to contact the company's Disaster Management Team Coordinators or Chairpersons with a damage assessment and actions taken report, and act upon instructions received.
6. If all documented attempts to communicate with Coordinators and Chairpersons have failed, the Office Leader or designate is authorized to initiate reasonable and prudent responses necessary to minimize potential:
    1. Injuries to personnel and customers.
    2. Damage to facilities.
    3. Loss of assets and records.

14.5 The critical functions of the Office are to be serviced before the performance of any other task. The critical Office functions are:

1. Employee, customer, facility, assets and records security.
2. Cash transfer and storage tasks, and facility security.
3. Acceptance of cash and negotiable documents as payment to accounts or for merchandise.
4. On-site delivery of merchandise.
5. Inventory and dual custody control.

14.6 The accessory functions of the Office are to be addressed only after all critical functions have been performed. The accessory functions of the Office are:

1. Courier service for cash and documents.
2. Account reconciliation and charge account balances.
3. Delivery and storage of merchandise.
4. Opening new customer accounts.

14.7 Description of **Office Leader's** duties and responsibilities during a disaster:

1. Ultimately responsible for overall Office operations, including all personnel, customers, facilities and assets.
2. Office Security Officer, Office Compliance Officer and Office Disaster Management Team Coordinator; interior and exterior Office physical security and appearance.
3. Ensure adequate supervision for all personnel and functions while absent from the company or unavailable for contact; operational quality control.
4. Respond to and comply with all regulations, policies and procedures regarding Office operations; prepare reports as required.

5. Customer service and public relations; resolving customer complaints; approve unusual or unique transaction when no other person has that immediate authority; provide information to supervisor for media relations and all requests for interviews from the press, radio and television.
6. Provide appropriate members of the Disaster Management Team with accurate and timely information updates regarding the Office's recovery efforts.
7. Other duties and responsibilities as required.

14.8 Description of **Assistant Office Leader's** duties and responsibilities:

1. Perform all duties and responsibilities of the Office Leader in his/her absence or because of his/her unavailability.
2. Ensure dual custody requirements are maintained for all functions; maintain key, employee information log for dual custody assignments; enforce employee and functional security procedures, Office opening and closing procedures.
3. Manage day-to-day operational functions, and directly supervise all staff personnel.
4. Other duties and responsibilities as required.

14.9 If a disaster occurs during working hours, the staff will evacuate the facility and assemble at:

PRIMARY:

Location:

SECONDARY:

Location:

A diagram of the facility and designated emergency staging areas should be provided at the end of this section.

14.10 If the Office cannot be used, Office operations will immediately shift to these alternate sites:

PRIMARY:

Location:

Address:

City:

State: ZIP:

Phone:

FAX:

SECONDARY:

Location:

Address:

City:

State: ZIP:

Phone:

FAX:

14.11 If the Office is still operable, this checklist describes the functions or sections upon which you will concentrate recovery efforts, and in what order. Before opening the Office:

1. Assess safety considerations for employees and customers.
2. Coordinate with emergency services agencies, if necessary.
3. Conduct a damage assessment of the building; determine levels of operation or restoration time for electricity, telephones, water, computer systems.
4. Ensure all areas of responsibility are staffed.
5. Ensure equipment and supplies are available.
6. Arrange for the safe relocation of all records and equipment, if necessary:
    - Actual customer transactions in progress.
    - Microfilm.
    - Daily balance records.
    - G/L reports.
    - Protect-a-graph equipment.
    - Supply of official checks and other negotiable documents.

14.12 If it is safe to open the Office, re-establish:

1. Employee, customer, facility, assets and records security.
2. Cash transfer and storage tasks, and facility security.
3. Install data communications equipment.
4. On-site delivery of merchandise.
5. Inventory and dual custody control.

14.13 The Office requires these logistical factors to be available to perform critical functions:

- Square feet: ______________________________
- Maximum number of employees: __________
- Maximum number of customers: __________

Identify special relocation needs in the event the facility is unable to support Office operations are:

____
____
____
____
____
____
____
____
____
____
____

14.14 A listing of all emergency services, personnel and equipment available to this Office is located in Section 9.0.

14.15 Additional guidelines to assist disaster recovery efforts for this Office are:
____

Written operations procedures:
____

Location stored:
____

Container description:
____

14.16 All service agreements and vendor information are available by contacting:

____
____

14.17 Additional office supplies, emergency equipment and survival supplies to assist disaster recovery efforts for this Office are:

- Emergency medical supplies available.

- Location stored.
- Container description
- An appropriate supply of the following forms is to be maintained:

____

____

____

____

____

____

____

____

____

____

____

____

____

____

PERSON COMPLETING:

TITLE:

DATE: FORWARDED TO: _______________

15.0 DEPARTMENT RECOVERY PLAN – Data Processing

Department: **Data Processing**

Address: ______________________________

City: ______________________________

State: ________ ZIP: ______________

Phone: ______________________________

FAX: ______________________________

Department Leader: ______________________________

Title: ______________________________

Assistant Department Leader: ______________________________

Title: ______________________________

15.1 Maintaining contact with members of the company's Disaster Management Team during a disaster is critical to a successful Department recovery effort. Usual business phone numbers are listed below, and these numbers should be used for all primary contact with Team members.

Disaster Management Chairpersons:

Name: ______________________________ Phone: __________

Emergency Phone: ______________________________

Cellular Phone: ______________________________

Name: ______________________________ Phone: __________

Emergency Phone: ______________________________

Cellular Phone: ______________________________

Disaster Management Coordinators:

Name: ______________________________ Phone: __________

Emergency Phone: ______________________________

Cellular Phone: ______________________________

Facilities/Equipment/Supplies / Transportation/Telecommunications:

Name: ______________________________ Phone: ______________

Emergency Phone: ______________________

Cellular Phone: ______________________

Other Department Leaders:

Name: ______________________________ Phone: ______________

Emergency Phone: ______________________

Cellular Phone: ______________________

Name: ______________________________ Phone: ______________

Emergency Phone: ______________________

Cellular Phone: ______________________

Name: ______________________________ Phone: ______________

Emergency Phone: ______________________

Cellular Phone: ______________________

Name: ______________________________ Phone: ______________

Emergency Phone: ______________________

Cellular Phone: ______________________

Name: ______________________________ Phone: ______________

Emergency Phone: ______________________

Cellular Phone: ______________________

15.2 In the event that normal phone lines are not functional, alternate communications may be available by using public phones (Bell system only). The public phones most readily accessible by Department personnel are:

Public Phone # 1: ______________________________

Location: ______________________________

Public Phone # 2: ______________________________

Location: ______________________________

Public Phone # 3: ______________________________

Location: ______________________________

15.3 The company has established a hotline phone number for emergency use by all employees. Use of this special number is restricted to disaster recovery efforts and emergency notifications only, and is not to be used for any other purpose.

Disaster Recovery Hotline number for employees only: ____________________

Security Alarm Company: ____

15.4 The Department Leader or designate is to immediately take the following actions if a disaster occurs:

1. Assess any injury or damage to employees, customers, and facilities.
2. Temporarily close and secure the facility, if necessary.
3. Contact appropriate emergency services, if necessary.
4. Begin documenting the effects of the disaster and actions taken; secure all assets and records.
5. First attempt to contact the company's Disaster Management Team Coordinators or Chairpersons with a damage assessment and actions taken report, and act upon instructions received.
6. If all documented attempts to communicate with Coordinators and Chairpersons have failed, the Department Leader or designate is authorized to initiate reasonable and prudent responses necessary to minimize potential:
    - Injuries to personnel and customers.
    - Damage to facilities.
    - Loss of assets and records.

15.5 The critical functions of the Department are to be serviced before the performance of any other task. The critical Department functions are:

1. Administrative operations.
2. Business operations.
3. Board of Directors.
4. Media relations.
5. Security.
6. Stock/shareholder relations.

15.6 The accessory functions of the Department are to be addressed only after all critical functions have been performed. The accessory functions of the Department are:

1. Credit Administration.
2. Personnel.
3. Accounting / Auditing.
4. Maintain custody of important company records.
5. Insurance policies and Board-related matters.

15.7 Description of Department Leader's duties and responsibilities during a disaster:

1. Ultimately responsible for overall Department operations, including all personnel, customers, facilities and assets.
2. Department Security Officer, Department Compliance Officer and Department Disaster Management Team Coordinator; interior and exterior Department physical security and appearance.
3. Ensure adequate supervision for all personnel and functions while absent from the company or unavailable for contact; operational quality control.
4. Respond to and comply with all regulations, policies and procedures regarding Department operations; prepare reports as required.
5. Customer service and public relations; resolving customer complaints; approve unusual or unique transaction when no other person has that immediate authority; provide information to supervisor for media relations and all requests for interviews from the press, radio and television.
6. Provide appropriate members of the Disaster Management Team with accurate and timely information updates regarding the Department's recovery efforts.
7. Other duties and responsibilities as required.

15.8 Description of Assistant Department Leader's duties and responsibilities:

1. Perform all duties and responsibilities of the Department Leader in his/her absence or because of his/her unavailability.
2. Ensure dual custody requirements are maintained for all functions; maintain key, employee information log for dual custody assignments; enforce employee and functional security procedures, Department opening and closing procedures.
3. Manage day-to-day operational functions, and directly supervise all staff personnel.
4. Other duties and responsibilities as required.

15.9 If a disaster occurs during working hours, the staff will evacuate the facility and assemble at:

PRIMARY: ____

Location: ____

SECONDARY: ____

Location: ____

A diagram of the facility and designated emergency staging areas is located at the end of this section.

15.10 If the Department is not able to be used, Department operations will immediately shift to these alternate sites:

PRIMARY: ______________________________

Location: ______________________________

Address: ______________________________

City: ______________________________

State: ______________ ZIP: ______________

Phone: ______________________________

FAX: ______________________________

SECONDARY: ______________________________

Location: ______________________________

Address: ______________________________

City: ______________________________

State: ______________ ZIP: ______________

Phone: ______________________________

FAX: ______________________________

15.11 If the Department is still operable, this checklist describes the functions or sections upon which you will concentrate recovery efforts, and in what order. Before opening the Department:

1. Assess safety considerations for employees and customers.
2. Coordinate with emergency services agencies, if necessary.
3. Conduct a damage assessment of the building; determine levels of operation for restoration time for electricity, telephones, water, computer system.
4. Ensure all areas of responsibility are staffed.
5. Ensure equipment and supplies are available.
6. Arrange for the safe relocation of all records and equipment, if necessary.

15.12 If it is safe to open the Department, re-establish:

- Employee, customer, facility, assets and records security.
- Contact with the Board of Directors.
- Contact with media representatives.
- Corporate files and financial records.
- Personnel and fixed asset records.
- Accounting records.
- Insurance policies and documentation.

15.13 The Department requires these logistical factors to be available to perform critical functions:

Square feet: __________

Maximum number of employees: __________

Maximum number of customers: __________

Special relocation needs in the event the facility is unable to support Department operations are:

____

____

15.14 A listing of all emergency services, personnel and equipment available to this Department is located in Section 13.0.

Additional guidelines to assist disaster recovery efforts for this Department are:

Written operations procedures: ____________________________

Location stored: ____________________________

Container description: ____________________________

15.15 All service agreements and vendor information are available by contacting:

____

____

15.16 Additional office supplies, emergency equipment and survival supplies to assist disaster recovery efforts for this Department are:

Emergency medical supplies available:

____

____

Location stored:

____

____

Container description:

____

____

An appropriate supply of the following forms is to be maintained:

PERSON COMPLETING: ____________________________

TITLE: ____________________________

DATE: ______________ FORWARDED TO: ______________

16.0 DEPARTMENT RECOVERY PLAN - General

Department: ______________________________

Address: ______________________________

City: ______________________________

State: ________ ZIP: ____________

Phone: ______________________

FAX: ______________________

Department Leader: ______________________________

Title: ______________________________

Assistant Department Leader: ______________________________

Title: ______________________________

16.1 Maintaining contact with members of the company's Disaster Management Team during a disaster is critical to a successful Department recovery effort. Usual business phone numbers are listed below, and these numbers should be used for all primary contact with Team members.

Disaster Management Chairpersons:

Name: ______________________ Phone: ____________

Emergency Phone: __________________

Cellular Phone: __________________

Name: ______________________ Phone: ____________

Emergency Phone: __________________

Cellular Phone: __________________

Disaster Management Coordinators:

Name: ______________________ Phone: ____________

Emergency Phone: __________________

Cellular Phone: __________________

Facilities/Equipment/Supplies / Transportation/Telecommunications:

Name: ______________________ Phone: ______________

Emergency Phone: ______________________

Cellular Phone: ______________________

Other Department Leaders:

Name: ______________________ Phone: ______________

Emergency Phone: ______________________

Cellular Phone: ______________________

Name: ______________________ Phone: ______________

Emergency Phone: ______________________

Cellular Phone: ______________________

Name: ______________________ Phone: ______________

Emergency Phone: ______________________

Cellular Phone: ______________________

Name: ______________________ Phone: ______________

Emergency Phone: ______________________

Cellular Phone: ______________________

Name: ______________________ Phone: ______________

Emergency Phone: ______________________

Cellular Phone: ______________________

16.2 In the event that normal phone lines are not functional, alternate communications may be available by using public phones (Bell system only). The public phones most readily accessible by Department personnel are:

Public Phone # 1: ______________________

Location: ______________________

Public Phone # 2: ______________________

Location: ______________________

Public Phone # 3: ______________________

Location: ______________________

16.3 The company has established a hotline phone number for emergency use by all employees. Use of this special number is restricted to disaster recovery efforts and emergency notifications only, and is not to be used for any other purpose.

Disaster Recovery Hotline number for employees only: ______________

Security Alarm Company: ______________________

16.4 The Department Leader or designate is to immediately take the following actions if a disaster occurs:

1. Assess any injury or damage to employees, customers, and facilities.
2. Temporarily close and secure the facility, if necessary.
3. Contact appropriate emergency services, if necessary.
4. Begin documenting the effects of the disaster and actions taken; secure all assets and records.
5. First attempt to contact the company's Disaster Management Team Coordinators or Chairpersons with a damage assessment and actions taken report, and act upon instructions received.
6. If all documented attempts to communicate with Coordinators and Chairpersons have failed, the Department Leader or designate is authorized to initiate reasonable and prudent responses necessary to minimize potential:
    - Injuries to personnel and customers.
    - Damage to facilities.
    - Loss of assets and records.

16.5 The critical functions of the Department are to be serviced before the performance of any other task. The critical Department functions are:

1. Administrative operations.
2. Business operations.
3. Board of Directors.
4. Media relations.
5. Security.
6. Stock/shareholder relations.

16.6 The accessory functions of the Department are to be addressed only after all critical functions have been performed. The accessory functions of the Department are:

1. Credit Administration.
2. Personnel.
3. Accounting / Auditing.
4. Maintain custody of important company records.
5. Insurance policies and Board-related matters.

16.7 Description of Department Leader's duties and responsibilities during a disaster:

1. Ultimately responsible for overall Department operations, including all personnel, customers, facilities and assets.
2. Department Security Officer, Department Compliance Officer and Department Disaster Management Team Coordinator; interior and exterior Department physical security and appearance.
3. Ensure adequate supervision for all personnel and functions while absent from the company or unavailable for contact; operational quality control.
4. Respond to and comply with all regulations, policies and procedures regarding Department operations; prepare reports as required.
5. Customer service and public relations; resolving customer complaints; approve unusual or unique transaction when no other person has that immediate authority; provide information to supervisor for media relations and all requests for interviews from the press, radio and television.
6. Provide appropriate members of the Disaster Management Team with accurate and timely information updates regarding the Department's recovery efforts.
7. Other duties and responsibilities as required.

16.8 Description of Assistant Department Leader's duties and responsibilities:

1. Perform all duties and responsibilities of the Department Leader in his/her absence or because of his/her unavailability.
2. Ensure dual custody requirements are maintained for all functions; maintain key, employee information log for dual custody assignments; enforce employee and functional security procedures, Department opening and closing procedures.
3. Manage day-to-day operational functions, and directly supervise all staff personnel.
4. Other duties and responsibilities as required.

16.9 If a disaster occurs during working hours, the staff will evacuate the facility and assemble at:

PRIMARY: ______________________________

Location: ______________________________

PRIMARY: ______________________________

Location: ______________________________

A diagram of the facility and designated emergency staging areas is located at the end of this section.

16.10 If the Department is not able to be used, Department operations will immediately shift to these alternate sites:

PRIMARY: ______________________________

Location: ______________________________

Address: ______________________________

City: ______________________________

State: ______________ ZIP: ______________

Phone: ______________________________

FAX: ______________________________

SECONDARY: ______________________________

Location: ______________________________

Address: ______________________________

City: ______________________________

State: ______________ ZIP: ______________

Phone: ______________________________

FAX: ______________________________

16.11 If the Department is still operable, this checklist describes the functions or sections upon which you will concentrate recovery efforts, and in what order. Before opening the Department:

1. Assess safety considerations for employees and customers.
2. Coordinate with emergency services agencies, if necessary.
3. Conduct a damage assessment of the building; determine levels of operation or restoration time for electricity, telephones, water, computer systems.
4. Ensure all areas of responsibility are staffed.
5. Ensure equipment and supplies are available.
6. Arrange for the safe relocation of all records and equipment, if necessary.

16.12 If it is safe to open the Department, re-establish:

1. Employee, customer, facility, assets and records security.
2. Contact with the Board of Directors.
3. Contact with media representatives.
4. Corporate files and financial records.
5. Personnel and fixed asset records.
6. Accounting records.
7. Insurance policies and documentation.

16.13 The Department requires these logistical factors to be available to perform critical functions:

Square feet: __________

Maximum number of employees: __________

Maximum number of customers: __________

Special relocation needs in the event the facility is unable to support Department operations are:

16.14 A listing of all emergency services, personnel and equipment available to this Department is located in Section 9.0.

16.15 Additional guidelines to assist disaster recovery efforts for this Department are:

Written operations procedures: ______________________________

Location stored: ___________________________________

Container description: ______________________________

16.16 All service agreements and vendor information are available by contacting:

____

____

16.17 Additional office supplies, emergency equipment and survival supplies to assist disaster recovery efforts for this Department are:

Emergency medical supplies available:

____

____

Location stored:

____

____

Container description:

____

____

An appropriate supply of the following forms is to be maintained:

____

PERSON COMPLETING: ____

TITLE: ____

DATE: _________________ FORWARDED TO: ________________________

17.0 TRAINING PROGRAM

17.1 The company's Disaster Management Plan Training Program is to be developed and delivered in conjunction with the local, state and federal provisions regarding workplace safety and security training, and other appropriate regulations.

17.2 Training is to be delivered at least annually.

Each Department Leader is responsible for establishing, maintaining and documenting all training meetings in conjunction with the Disaster Management Team Coordinators.

17.3 Training is to include all personnel according to the following guidelines:

1. **Orientation**

   All new employees, regardless of position description or terms of employment, shall be oriented regarding the company's plan for the facility in which the employee works, and his/her role as assigned.

2. **Transfer/promotion**

   Any employee who has been transferred to a new position or facility, or who has been promoted, shall be re-oriented regarding the company's plan for the facility in which the employee is to work, and his/her new role as assigned.

3. **Updating**

   Any change in policies, procedures, equipment, roles or personnel affecting the implementation of the Disaster Management Plan shall be addressed at the time the change occurs.

17.4 All employees shall be trained according to the following minimum standards:

1. Company's policies and procedures for responding to crimes of violence: robbery, kidnapping, extortion, bomb threats, threatening phone calls and other identified potential threats. This training shall include the company's philosophy, reporting requirements, individual duties and responsibilities, behavioral expectations and performance standards.
2. Demonstration of all electrical, water, gas and other appropriate shut-off devices, including the device's location, access requirements, physical operation, and individual duties and responsibilities for shutting these devices off during an emergency.
3. Demonstration of the facility's emergency staging areas, to be located an appropriate distance from the facility, in an area clearly visible from the facility and the surrounding area.

4. Evacuation and building plans, and demonstration of all procedures to be used during an emergency, including individual duties and responsibilities, reporting requirements, behavioral expectations and performance standards. This may also include a timed mock exercise to test the effectiveness of the evacuation procedures.

   If a timed mock evacuation exercise is to be delivered, prior notification shall be given to and approval received from the:

   1. Disaster Management Team Coordinators.
   2. Local law enforcement agency.
   3. Fire department, if appropriate.

5. Alert procedures for notifying personnel of an emergency, including the stages of alert and individual duties and responsibilities for all phases of the recovery effort.

6. Location, purpose and use of emergency equipment, fire extinguishers, first aid kits and other survival tools and supplies stored at the facility.

7. Basic first aid techniques, if appropriate.

17.5 After the Disaster Management Plan has been approved by the Board of Directors, the following training meetings may be delivered:

1. **Board of Directors**:

   An orientation meeting to discuss the Board's responsibilities regarding implementation of the Disaster Management Plan, and the regulations issued by or in relation to local, state or federal agencies, as they affect the implementation of the company's Disaster Management Plan.

   This training may be delivered once to all Board members, and at the time a new member is appointed. An annual update meeting may also be held before the Board meeting addressing the re-certification of the Disaster Management Plan. The Board Secretary will reflect the training meeting in the Board minutes.

2. **Executives:**

   An orientation meeting to discuss the executives' responsibilities regarding implementation of the Disaster Management Plan, and the regulations issued by or in relation to local, state or federal agencies, as they affect the implementation of the company's Disaster Management Plan.

   Training will be delivered once to all executive officers, and at the time a new executive officer is appointed. An update meeting will also be held at least annually, before the Board meeting addressing the re-certification of the Disaster Management Plan. The Board Secretary will reflect the training meeting in the Board minutes.

3. **Other Officers, Department Leaders and Office Managers**:

   An orientation meeting to discuss these groups' responsibilities regarding implementation of the Disaster Management Plan, and the regulations issued by or in relation to local, state or federal agencies.

   This training will be delivered once to all other officers, department leaders and office managers, and at the time a new manager/officer is appointed. An update meeting will also be held at least annually, before the Board meeting addressing there-certification of the Disaster Management Plan. The Board Secretary will reflect the training meeting in the Board minutes.

Notes

18.0 UPDATE INFORMATION

The company has updated information, added to this plan, and distributed new pages/copies on the following dates:

SECTION#	DATE	EMPLOYEE	COMMENTS

(continue on back)

SECTION#	DATE	EMPLOYEE	COMMENTS

19.0 SAMPLE FORMS

19.1 The forms in this manual are to be used in conjunction with the sections indicating such use.

19.2 All forms are to be completed, signed and delivered as specified.

19.3 The forms contained in this section are the only forms authorized for use by company personnel. All new forms affecting the Disaster Management Plan are to be approved by the Disaster Management Team Coordinators before implementation.

19.4 The forms contained in this section are:

1. Disaster Recovery Log	Exhibit 1
2. Chronological Log of Events	Exhibit 2
3. Testing of Solutions Documentation	Exhibit 3
4. Employee Personal Profile	Exhibit 4
5. <u>Department Checklists and Worksheets:</u>	
RECOVERY WORKSHEET FOR:	Exhibit 5
RESTORATION WORKSHEET FOR:	Exhibit 6
RESUMPTION WORKSHEET FOR:	Exhibit 7
RECONSTRUCTION WORKSHEET FOR:	Exhibit 8

[This page intentionally left blank]

DISASTER RECOVERY LOG

DATE: __________ TIME STARTED: ________ TIME ENDED: __________

OFFICE/DEPT: ____

OPERATOR: ____

OFFICE/DEPT CONTACT: _______________________________________

PHONE: ( )/____________________ FAX: ( )/____________________

LOCATION: ____

ALTERNATE: ____

PHONE: ( )/____________________ FAX: ( )/____________________

LOCATION: ____

RECOVERY SITE: ____

RECOVERY SITE: ____

RECOVERY SITE: ____

Notes ____

EXHIBIT 1

[This page intentionally left blank]

CHRONOLOGICAL LOG OF EVENTS

DATE: __________ TIME STARTED: ________ TIME ENDED: ________

LOCATION: ____

MANAGER IN CHARGE: _________________ OPERATOR: _______________

Time	Person	Activity/request	Disposition	Time

EXHIBIT 2

[This page intentionally left blank]

TESTING OF SOLUTIONS DOCUMENTATION

On ________________________ the Disaster Management Plan solutions to the scenario indicated below were tested with the following results:

SCENARIO:

____

____

____

____

____

____

Test conducted by: ____

Location/Department: ____

Date: ___________________________ Time: _______________________

Results:

____

____

____

____

____

____

____

____

____

____

EXHIBIT 3

[This page intentionally left blank]

EMPLOYEE PERSONAL PROFILE – page 1

The information requested in this questionnaire is voluntary and confidential, and is not to be used for any purpose other than during an actual emergency. The contents of this questionnaire shall be kept in a sealed envelope in a secure area, and shall not be opened unless in the case of an actual emergency. The contents of this questionnaire and a photograph shall be updated annually during your performance evaluation.

PERSONAL IDENTIFYING INFORMATION:

Employee:

____

Nicknames or other names used:

____

Employment classification:

____

Employment location:

____

Permanent residence:

____

_____________________________________ Phone: _________

Secondary residence:

____

_____________________________________ Phone: _________

Other employment, if applicable:

____

Date of birth: / / / Place of birth:

____

Name of hospital: _______________ Mother's name: ___________

Race: _______ Sex: _______ Complexion: ____________________

Height: _____ Weight: ____ Hair color: _____ Eye color: _____

Scars/marks/tattoos: _____________________________

Hobbies: ______________________________________

Are your fingerprints on file with the company? ____________

Is a current photograph on file with the company? ___________

EXHIBIT 4-1

EMPLOYEE PERSONAL PROFILE - page 2

FAMILY AND EMERGENCY NOTIFICATION INFORMATION:

Marital status: ________________ Anniversary date: __________

Name of spouse/roommate: ____________ Nickname: ___________

Number of children: _______ Children's names are:

____________________________________ Birth date: ____________

____________________________________ Birth date: ____________

____________________________________ Birth date: ____________

____________________________________ Birth date: ____________

PERSONS TO CONTACT IN CASE OF EMERGENCY:

Name: __________________________________ Phone: _____________

Address: _______________________________ Relationship: __________

Name: __________________________________ Phone: _____________

Address: _______________________________ Relationship: __________

Name: __________________________________ Phone: _____________

Address: _______________________________ Relationship: __________

IMMEDIATE CLOSE RELATIVES:

Name: _______________________________ Relationship: ____________

Address: _______________________________ Phone: ________________

Name: _______________________________ Relationship: ____________

Address: _______________________________ Phone: ________________

Name: _______________________________ Relationship: ____________

Address: _______________________________ Phone: ________________

EXHIBIT 4 - 2

EMPLOYEE PERSONAL PROFILE - page 3

OTHER PERSONS LIVING OR WORKING IN HOUSEHOLD:

Name: ______________________ Relationship: ____________

Address: ______________________ Phone: ______________

Name: ______________________ Relationship: ____________

Address: ______________________ Phone: ______________

MOTOR VEHICLES USED BY FAMILY

Year: ___ Make: __________ Model: ________ Color: ________

License: ____________ Driven by: ______________________

Year: ___ Make: __________ Model: ________ Color: ________

License: ____________ Driven by: ______________________

MEDICAL INFORMATION

Blood type: _____ Allergic to: ______________________________

Medical condition requiring treatment or medication:

____

____

Medication required: ______________________ Dosage: ________

Personal physician: ______________________ Phone: ________

Address: ____

AUTHORIZATION FOR THIS PHYSICIAN TO RELEASE CONFIDENTIAL INFORMATION IN THE EVENT OF AN EMERGENCY SITUATION REQUIRING TREATMENT

Signature: ________________________ Date: ______________

EXHIBIT 4-3

EMPLOYEE PERSONAL PROFILE - page 4

EMPLOYEE:

After completing this form, place the form into the envelope provided and seal the envelope. Write your name (last, first, middle initial) and Social Security number on the face of the envelope. Deliver the sealed envelope to your reviewer.

Please include any other information you feel is necessary on a separate page, to be included with this profile.

____

____

____

____

REVIEWER:

Ensure that the employee's envelope is securely sealed and completed as indicated. Initial and date the face of the envelope, and store in a secure area.

____

____

____

____

____

____

____

____

EXHIBIT 4 - 4

RECOVERY WORKSHEET

For: ________________________________

Following the onset of an emergency or disaster requiring the activation of the company's Disaster Management Plan, the following procedures must be implemented to recover/stabilize operations:

1. Follow emergency instructions for specific position descriptions, addressing employee and customer safety first.
2. Assess any injury to employees and customers, and damage to facilities.
3. Temporarily close and secure the facility, if necessary.
4. Contact appropriate emergency services, if necessary.
5. Begin documenting the effects of the disaster and actions taken; secure all assets and records.
6. First attempt to contact the Disaster Management Team Coordinators with a damage assessment and actions taken report, and act upon instructions received. If the Coordinators are unavailable, contact Chairpersons for instructions.
7. If all documented attempts to communicate with Coordinators and Chairpersons have failed, the Department/Office Leader or designate is authorized to initiate reasonable and prudent responses necessary to minimize potential:
    1. Injuries to personnel and customers;
    2. Damage to facilities; and
    3. Loss of assets and records.

NOTES:

____

____

____

EXHIBIT 5

[This page intentionally left blank]

RESTORATION WORKSHEET

For: ____

Following the recovery/stabilization of operations, the following procedures must be implemented to restore minimum levels of service:

1. Based upon the actual results of the disaster, review and temporarily revise the disaster response plan appropriately.
2. Assign the performance of critical tasks to designated employees, and initiate and maintain reporting structure and requirements.
3. Conduct a damage assessment of all equipment and supplies. Arrange for the transportation of replacement equipment and supplies, if appropriate.
4. Verify that all necessary equipment and supplies are available to perform critical tasks.
5. Retrieve data files, vital records, and other critical work.
6. Provide for physical site security, and institute other security policies and procedures.
7. Clear the work areas and provide temporary repairs as appropriate.
8. Continue documenting the effects of the disaster, and all actions taken to restore services. Photograph the impact areas if possible.
9. ____
10. ____

EXHIBIT 6

[This page intentionally left blank]

RESUMPTION WORKSHEET

For: ____

Following the restoration of basic operations, the following procedures must be implemented to resume normal levels of service:

1. Verify that basic operations have been fully restored, and notify the Disaster Management Team Coordinators of the completed restoration.
2. Determine personnel, equipment and supply requirements to complete the restoration of normal services.
3. Replace or restore equipment as appropriate.
4. Continue documenting the effects of the disaster, and all actions taken to restore services.
5.
6.
7.
8.
9.
10.
11.
12.

EXHIBIT 7

[This page intentionally left blank]

RECONSTRUCTION WORKSHEET

For: ____

Following the resumption of normal operations, the following procedures must be implemented to reconstruct this department to the level of operation existing prior to the emergency:

1. ____
2. ____
3. ____
4. ____
5. ____
6. ____
7. ____
8. ____
9. ____
10. ____
11. ____
12. ____

EXHIBIT 8

[This page intentionally left blank]

20.0 RESOLUTIONS FOR THE DISASTER MANAGEMENT PLAN

20.1 The purpose of this section is to insure the continuity of management and business operations in the event of a declared emergency, by providing for emergency operating policies and procedures to be implemented by the surviving staff.

20.2 In the event of an emergency declared by the President of the United States, the Governor of this State, or another competent and authorized public authority, or the person designated to perform his/her functions, or the President of the company or his/her designate, the officers and employees of the company shall continue to conduct the affairs of the company, except as otherwise provided in the Disaster Management Plan. The conducting of the affairs of the company shall be subject to such guidance from the Board of Directors as may be available, except as to matters which, by statute or by-laws, require specific approval of the Board, and subject to conformance with any governmental directives during the emergency.

20.3 The Board of Directors shall have the power to delegate and prescribe an officer's powers and duties to any other officer or director on a temporary basis, when such officer is unavailable to, or incapable of, performing such duties; or if such officer refuses to act according to the policies and procedures set forth in this resolution.

20.4 In the event of a state of disaster or emergency in the manner described in Section 20.2 of this resolution of sufficient severity to prevent the conduct and management of the affairs and business of the company by its officers and directors, as set forth by this resolution, and two or more available members of the then-incumbent Disaster management Team or Disaster Management Executive Committee are present, they shall constitute a quorum of that Disaster Management Executive Committee for the full conduct and management of the affairs and business of the company, in accordance with the provisions of the company's by-laws.

20.5 In addition, such Disaster Management Team or Disaster Management Executive Committee shall be empowered to exercise all the powers normally reserved for all other committees, including the Board of Directors, according to the company's by-laws, although those committees may function in routine operation to the extent possible, with the guidance of the Disaster Management Team. No officer, director or employee acting according to the emergency by-laws or resolutions shall be personally liable except for acts of willful misconduct.

20.6 This Disaster Management Plan resolution shall be subject to the approval of and implementation by the Board of Directors, and periodic review by that Board, and shall, upon approval by the Board, immediately become by-laws to be enacted during a declared emergency. Any provisions of other by-laws and resolutions that are contrary to the provisions of this section shall be suspended until it shall be determined by an interim Disaster Management Team or Disaster Management Executive Committee, or the company's Board of Directors acting under this section, that it shall be to the advantage of the company to resume the ordinary conduct and management of its affairs and business.

20.7 It is resolved that, upon the declaration of disaster or emergency in the manner described in Section 20.2 of this resolution, the Chief Executive Officer of the company cannot be located by the then-acting Main Office, or is unable to assume or continue normal executive duties, then the authority, duties and responsibilities of the Chief Executive Officer shall, without further action by the Board of Directors, be automatically assumed by one of the following persons in the designated order:

1. ______________________________
2. ______________________________
3. ______________________________
4. ______________________________
5. ______________________________
6. ______________________________

20.8 Any one of the listed persons who assumes the authority, duties and responsibilities of the Chief Executive Officer shall continue to serve in that capacity until he/she resigns; or until two-thirds of the available officers of the company are available and decide, in writing, that he/she is unable to perform such duties; or until the elected Chief Executive Officer of the company or a person higher on the indicated list becomes available to perform the duties of the Chief Executive Officer of the company, in which case that person will assume the office.

20.9 It is further resolved that, upon the declaration of a disaster or emergency in the manner described in Section 20.2 of this resolution, the President of the company cannot be located by the then-acting Main Office, or is unable to assume or continue normal executive duties, then the authority, duties and responsibilities of the President shall, without further action by the Board of Directors, be automatically assumed by one of the persons on the list indicated in Section 20.7 in the designated order.

20.10 The person assuming the authority, duties and responsibilities of the President according to this resolution shall serve in this capacity until the elected President or person higher on the indicated list shall be able to function as President; or until he/she resigns; or is unable as determined by the Chief Executive Officer to perform the duties of his/her office. If this acting President is replaced, such replacement shall be by the next eligible and available person on the indicated list. The foregoing sections 20.8 and 20.9 shall be applicable only if the same person does not serve as the President and as the Chief Executive Officer.

20.11 It is also resolved that any person or entity having business with the company, may accept a certification by any three officers that a specified person is acting as Chief Executive Officer or President of the company, according to this resolution. This person or entity accepting this certification may continue to consider it in force until notified in writing of a change, such notice of change to carry the signatures of the same three officers of the company.

20.12 The offices of the company at which its business shall be conducted during a declared disaster or emergency shall be:

The Main Office

Address: ______________________________

City: ______________________________

State: __________ ZIP: ____________________

Phone: ______________________________

Plus any of its facilities or other legally authorized locations that may be leased or acquired by the company for conducting its business.

20.13 During an emergency or disaster that causes any normal place of business of the company to cease operations, the business normally conducted at that location shall be relocated elsewhere in suitable quarters, in addition to or in lieu of the locations referred to in this resolution. These locations may be designated by the Board of Directors, the Disaster Management Team or Disaster Management Executive Committee, or by such other persons as are then, according to the resolutions adopted from time to time by the Board of Directors to deal with the exercise of authority in the time of such emergency, and the conduct of the affairs and business of the company as set forth in the company's by-laws.

20.14 Any temporarily relocated place of business of the company shall be returned to its legally authorized location by the Disaster Management Team or Disaster Management Executive Committee as soon as it is practical, and the temporary place of business shall be discontinued.

20.15 It is resolved that, in the event of a disaster or emergency declared as described in Section 20.2 of this resolution, or because of war or warlike damage or disaster, the Main Office shall serve as headquarters for conducting all affairs and business of the company.

20.16 It is further resolved that, in the event of a disaster or emergency declared as described in Section 20.2 of this resolution, or because of war or warlike damage or disaster, the Main Office is temporarily unable to continue its functions, the following locations shall be designated by the Disaster Management Team or Disaster Management Executive Committee as the acting Main Office of the company, automatically and without further action of the Board of Directors, in the listed order:

Acting Main Office #2

Address: ______________________________

City: ______________________________

State: ______________ ZIP: ______________

Phone: ______________________________

Acting Main Office #3

Address: ______________________________

City: ______________________________

State: ______________ ZIP: ______________

Phone: ______________________________

Acting Main Office #4

Address: ______________________________

City: ______________________________

State: ______________ ZIP: ______________

Phone: ______________________________

20.17 These alternate offices of the company shall function as the Main Office until and unless relieved of this responsibility by the Chief Executive Officer of the company, acting according to the written instructions of the Board of Directors, in the form of a resolution.

20.18 The following temporary emergency powers are granted to the Chief Executive Officer or designate:

1. To close any office, department, facility or function;
2. To establish accounts in the name of the company, to deposit into and withdraw funds from any and all company accounts, and to sign checks and other negotiable documents drawn upon those accounts;
3. To borrow funds from any governmental agency or other financial company from time to time, in the company's name;

4. To sell any U.S. Government obligation or other investment securities owned by the company, and to pledge, mortgage or hypothecate any such securities or other properties as may be required by a lender in support of any borrowing contemplated under Paragraph 2 hereof or in support of any other company obligation;
5. To execute such promissory note or notes or other documents or agreements as may be required by a lender in support of any borrowing contemplated under Paragraph 3 hereof, or to carry out any action as may be necessary to fully perform the provisions of any such notes, documents or agreements, in the name of the company; and
6. To delegate in writing the authority vested in him/her under this resolution to any other officer or officers or employee or employees within such limits and under such conditions as he/she may specify, and such officer or officers, or employee or employees, shall thereupon have the power to act within the authority so delegated.

20.19 The authorization of this resolution is granted during the Board meeting held as evidenced below.

______________________________________ ______________________

Secretary ____Date: _______________

21.0 DISASTER MANAGEMENT EXECUTIVE COMMITTEE - APPOINTMENT

21.1 The indicated persons have been selected to fulfill Disaster Management Executive Committee positions in the event an emergency activation is declared. The duties and responsibilities of this Disaster Management Executive Committee shall commence upon the actual declaration of emergency by the Chairpersons of the Disaster Management Team, and terminate upon personal notification by the Chairpersons or their designate(s).

21.2 The duties and responsibilities of the Disaster Management Executive Committee shall be limited to the term of the actual emergency, and this Disaster Management Executive Committee is hereby authorized to:

1. In cooperation with other members of the Disaster Management Executive Committee, provide for the succession of management in the event that the person normally assigned to that position is unable, for whatever reason, to perform the functions essential to that position to the satisfaction of Disaster Management Executive Committee.
2. In cooperation with other members of the Disaster Management Executive Committee, assign emergency powers as necessary to insure that the safety of personnel, customers, facilities, assets and records is not compromised.
3. In cooperation with other members of the Disaster Management Executive Committee, enact temporary policies and procedures as necessary during the emergency condition, when the Disaster Management Executive Committee determines that normal policies and procedures are ineffective.

21.3 The members of the company's Disaster Management Executive Committee are:

1. Co-Chairperson: ______________________________
2. Co-Chairperson: ______________________________
3. Co-Coordinator: ______________________________
4. Co-Coordinator: ______________________________
5. Department: ______________________________
6. Department Leader: ______________________________
7. Assistant Department Leader: ______________________________
8. Department: ______________________________
9. Department Leader: ______________________________
10. Assistant Department Leader: ______________________________
11. Department: ______________________________
12. Department Leader: ______________________________
13. Assistant Department Leader: ______________________________

14. Department: ____
15. Department Leader: _______________________________________
16. Assistant Department Leader: ______________________________
17. Department: ____
18. Department Leader: _______________________________________
19. Assistant Department Leader: ______________________________
20. Department: ____
21. Department Leader: _______________________________________
22. Assistant Department Leader: ______________________________

21.4 The authorization of this Disaster Management Executive Committee has been granted during the Board meeting indicated below, as evidenced by the seal and signature of the Secretary.

______________________________ ______________________________

Secretary Date

22.0 DISASTER MANAGEMENT EXECUTIVE COMMITTEE - ACKNOWLEDGMENT

22.1 You have agreed to fulfill a position on the Disaster Management Executive Committee, in the event an emergency activation of this plan is declared. The duties and responsibilities of this Disaster Management Executive Committee shall commence upon the actual declaration of emergency by the Chairpersons of the Disaster Management Team, and terminate upon personal notification by the Disaster Management Team Chairperson or their designate(s).

22.2 Your duties and responsibilities as a member of the Disaster Management Executive Committee shall be limited to the term of the actual emergency, and as a Disaster Management Executive Committee member, you are hereby authorized to:

1. In cooperation with other members of the Disaster Management Executive Committee, provide for the succession of management in the event that the person normally assigned to that position is unable, for whatever reason, to perform the functions essential to that position to the satisfaction of Disaster Management Executive Committee.
2. In cooperation with other members of the Disaster Management Executive Committee, assign emergency powers as necessary to insure that the safety of personnel, customers, facilities, assets and records is not compromised.
3. In cooperation with other members of the Disaster Management Executive Committee, enact temporary policies and procedures as necessary during the emergency condition, when the Disaster Management Executive Committee determines that normal policies and procedures are ineffective.

22.3 The members of the company's Disaster Management Executive Committee are:

1. Co-Chairperson: ______________________________

   Co-Chairperson: ______________________________
2. Co-Coordinator: ______________________________

   Co-Coordinator: ______________________________
3. Department: ______________________________

   Department Leader: ______________________________

   Assistant Department Leader: ______________________________
4. Department: ______________________________

   Department Leader: ______________________________

   Assistant Department Leader: ______________________________

5. Department: ______________________________
   Department Leader: ______________________________
   Assistant Department Leader: ______________________________
6. Department: ______________________________
   Department Leader: ______________________________
   Assistant Department Leader: ______________________________
7. Department: ______________________________
   Department Leader: ______________________________
   Assistant Department Leader: ______________________________

22.4 Your authorization as a member of this Disaster Management Executive Committee has been granted by the Board of Directors. Your signature below is simply an acknowledgment that you:

1. Accept these responsibilities.
2. Agree to attend all meetings and training events concerning the Disaster Management Plan.
3. Will remain up-to-date regarding the policies and procedures developed as a result of these meetings.
4. Agree to perform those duties to the best of your ability, and to notify the Chairpersons of the Disaster Management Team immediately in the event that you are no longer able, for whatever reason, to adequately fulfill these responsibilities.

______________________________ ______________________

Name Date

23.0 BOARD OF DIRECTORS' CERTIFICATION RESOLUTION

23.1 The Board of Directors of the company hereby declares that it has assigned the review of the contents of this Disaster Management Plan in its entirety, and has received a recommendation from the Disaster Management Team Coordinators and the Disaster Management Team Chairpersons regarding the adoption of this plan as the official plan for the company.

23.2 Pursuant to these guidelines, and upon the declaration of disaster in the manner set forth in this plan, the Disaster Management Team and Disaster Management Executive Committee are authorized, on an interim basis and without further authorization, to use any reasonable and necessary financial resources possessed by the company to successfully recover from the disaster.

23.3 The Board of Directors of the company hereby authorizes this Disaster Management Plan as the only such plan for the company, and authorizes the Disaster Management Team Chairpersons and their designate(s) to carry out all provisions of this plan. This authorization is granted during the Board meeting held as evidenced below.

______________________________ __________

Secretary Date

24.0 BOARD OF DIRECTORS' RE - CERTIFICATION RESOLUTION

24.1 The Board of Directors of the company hereby declares that it has reviewed the annual report concerning the Disaster Management Plan in its entirety, and has received a recommendation from the Disaster Management Team Coordinators and the Disaster Management Team Chairpersons regarding the re-certification of this plan as the continuing official plan for the company, subject to the revisions attached to this resolution.

24.2 Pursuant to these guidelines, and upon the declaration of a disaster in the manner set forth in this plan, the Disaster Management Team and Disaster Management Executive Committee are again authorized, on an interim basis and without further authorization, to use any reasonable and necessary financial resources possessed by the company to successfully recover from the disaster.

24.3 The Board of Directors of the company hereby authorizes this Disaster Management Plan as the only such plan for the company, and authorizes the Disaster Management Team Chairpersons and their designate(s) to carry out all provisions of this plan. This authorization is granted during the Board meeting held as evidenced below.

______________________________ ____________________

Secretary Date

[This page intentionally left blank]

25.0 COPIES OF RESOLUTIONS & SUCCESSION DOCUMENTATION

The resolution governing the approval of the company's Disaster Management Plan, all support documents, and a copy of the company's emergency succession shall be maintained in a safe deposit box or other dual custody container at the following locations:

1. Main Administrative Office:

Address: ______________________________
City: ______________________________
State: ______________ ZIP: ______________
Phone: ______________________
Box number: # __________________
Managers having keys:
1. ______________________________
2. ______________________________
3. ______________________________

2. ________________________

Address: ______________________________
City: ______________________________
State: ______________ ZIP: ______________
Phone: ______________________
Box number: # __________________
Managers having keys:
1. ______________________________
2. ______________________________
3. ______________________________

3. ________________________

Address: ______________________________
City: ______________________________
State: ______________ ZIP: ______________
Phone: ______________________
Box number: # __________________
Managers having keys:
1. ______________________________
2. ______________________________
3. ______________________________

26.0 CONFIDENTIAL INFORMATION

26.1 The listing of confidential information may be retained in a separate binder, but shall be stored with the Disaster Management Plan. This information is to be treated as confidential, and is to be used only in the event of an activation of this plan. This listing shall include, but is not limited to:

1. Emergency services agencies and community resources.
2. Equipment.
3. Facilities, if appropriate.
4. Personnel.
5. Vendors.

26.2 This listing is to be retained on a database, if available, and updated at least annually.

26.3 In the same binder is to be stored a copy of position descriptions for every position assigned within the company. These position descriptions are to be updated whenever a position has been created or changed significantly.

27.0 LEADER'S GUIDE: DISASTER MANAGEMENT AND BUSINESS RECOVERY PLAN

LEADER: Department or Office Leader

DISCUSSION TIME: 45 minutes

27.1 Overview

The company has developed and implemented a comprehensive disaster management and business recovery plan for all personnel, departments, offices, facilities and functions. Training the company's personnel to appropriately implement the plan is critical to the plan's success.

27.2 Objectives

- To inform all personnel about the company's Disaster Management Plan, and the policies and procedures regulating its implementation, on a company-wide scale.
- To advise all personnel about the company's Disaster Management Plan, and the policies and procedures regulating its implementation, as the plan affects specific functions, offices and departments.
- To train each employee to respond appropriately during a disaster, according to specific instructions designed for the employee's job classification.
- To suggest safety precautions for use in both a business and personal environment.

27.3 Discussion Materials

- Handout: Disaster Management Plan Training Handout
- Office or Department Disaster Recovery Plan script (Section 16.0)
- Emergency Staging Area and Facility Diagram
- Checklist

27.4 Meeting Preparation

- Conduct an interior and exterior inspection of the facility, noting potential hazards for discussion.
- Note locations of water, electrical and water connections, if appropriate. Do not attempt to test the switches or valves. Determine if any shut-off device requires a tool to operate: if so, determine if a tool is readily available.
- Verify compliance with local codes regarding electrical outlets, exit sign devices, hazardous materials storage standards, and other appropriate requirements. (Local fire marshal has a variety of inspection checklists.)

- Verify that the emergency staging areas are correctly designated on the plan diagram, and that the areas selected are free from inherent hazards (i.e., overhead electrical wires, fire plug, gas main, etc.)
- Verify the locations and numbers of the listed public telephones.
- Verify that lists containing information regarding the facility's personnel, emergency services and related agencies, equipment and appropriate records are current and available.
- Verify that first aid kits and fire extinguishers are available and functional.
- Verify that emergency supplies of food, water and survival equipment are available and stored appropriately.
- Verify that an appropriate inventory is maintained of all necessary forms and supplies.
- Designate appropriate employees to perform necessary functions during a disaster:
    - Evacuation of other employees and customers
    - Securing assets and records
    - Securing the facility and utilities
    - Record-keeping and report writing

27.5 Prepare for the meeting by

- Reading the handout to familiarize yourself with the information, and to insure that you understand the key elements of disaster management and business recovery.
- Discuss questions or concerns with your supervisor.
- Distributing the handout and any related materials to each employee in your office or department one (1) week before your meeting. Instruct employees to read the handout, note questions or comments directly on the copy, and to bring the handout to the meeting.
- Assembling all appropriate materials, displays and the handout to use as references during the meeting. Have extra copies of the handout available at the meeting.
- Preparing a meeting outline on a flip chart to help guide your presentation, if appropriate.
- Preparing a training roster (copy attached).

27.6 Conduct the Meeting

- Assign an employee to complete the roster and take notes during the meeting, and to deliver the signed roster and notes to you at the end of the meeting for your report. (A device is attached to this guide.)
- Introduce the purpose of the meeting, and the training program objectives. (Refer to your flip chart, if appropriate, throughout the presentation.)
- Refer to the handout, and; Define a disaster; discuss the types of disasters; the potential impact upon the community, the company and personally. Emphasize the importance of preparation.
- Discuss the primary goals of disaster management and business recovery planning, and how important the plan is to the company and its customers.
- Discuss the plan and emphasize the role and responsibilities of each team member and employee, in the event of a disaster.
- Discuss the facility's emergency staging areas and evacuation routes, and designate the employees responsible for evacuation.
- Discuss the facility's utilities (water, electricity and gas) and appropriate locations; emergency telephone numbers and their appropriate use; and the potential re-location of staff to an alternate work site, and describe those designated sites.
- Discuss public and media relations, and each employee's role in protecting the company's image.
- Discuss personal responsibilities and the potential impact upon business recovery, and the suggestions for personal preparation for a disaster.
- Ask that any employee having appropriate comments, concerns or questions contact you after the meeting, if an employee is not comfortable discussing these issues during the meeting.
- Distribute the training roster for employees' names and signatures.

27.7 After the meeting

- Retrieve the roster and notes from the note-taking employee.
- Using the notes and your personal observation, indicate in writing what issues need to be resolved and your suggestions for resolution.
- Complete the information requested on the attached memo, and return the memo, training roster, and your written report to the Disaster Management Team Coordinators.

[This page intentionally left blank]

MEMO

To: Disaster Management Team Coordinators

From:

Date:

Subject: Annual Training Meeting I conducted a site inspection for the ______________________ Office/Facility/Department on ________________________.

I reviewed the office/department/facility's recovery plan, Section 16.0 on ______________________.

I conducted a training meeting regarding the company's Disaster Management Plan on __________________. A copy of the training roster is attached.

Check one:

_________The office/facility/department is in compliance with all provisions of its disaster recovery plan.

_________The office/facility/department is not in compliance with all provisions of its disaster recovery plan.

Please contact me immediately to discuss the indicated items:

_______ **POLICY OR PROCEDURE:**

____

____

_______ **EQUIPMENT OR SUPPLIES:**

____

____

____

____

Name ______________________________Title: ____________________________

[This page intentionally left blank]

DISASTER MANAGEMENT & BUSINESS RECOVERY PLAN TRAINING

Training Roster Date: ___________________

Office/Department/Facility: ____________________________________

Printed Name	Signature

(page 1 of 2)

DISASTER MANAGEMENT & BUSINESS RECOVERY PLAN TRAINING

Printed Name	Signature

(page 2 of 2)

SEMINAR EVALUATION - DISASTER MANAGEMENT PLAN
ANNUAL TRAINING

LOCATION: ______________________ DATE: __________

YOUR NAME (Optional): ______________________

YOUR TITLE: ______________________

SEMINAR CONTENT AND ORGANIZATION	**Poor >>>>> Excellent**
1. Will you be able to apply this information in your position?	1 . 2 . 3 . 4 . 5
2. Was the seminar content well-organized?	1 . 2 . 3 . 4 . 5
3. Were the handout materials useful and informative?	1 . 2 . 3 . 4 . 5
4. Were the seminar topics adequately addressed?	1 . 2 . 3 . 4 . 5
SEMINAR INSTRUCTOR:	
5. Were questions answered to your satisfaction?	1 . 2 . 3 . 4 . 5
6. Was the instructor organized and well-prepared?	1 . 2 . 3 . 4 . 5
7. Was the instructor knowledgeable and up-to-date in the subject matter?	1 . 2 . 3 . 4 . 5
8. Did the instructor display enthusiasm for the topics and the class responses?	1 . 2 . 3 . 4 . 5
9. Were you encouraged to ask questions and express opinions?	1 . 2 . 3 . 4 . 5
10. Were the teaching aids, handout material and techniques effective?	1 . 2 . 3 . 4 . 5

COMMENTS: (On reverse, please):

(page 1 of 2)

Seminar Comments

(page 2 of 2)

DISASTER RECOVERY

Action Plan
400

[This page intentionally left blank]

ACTION PLAN

Table of Contents*

* The Action Plan Table of Contents references section page numbers (lower right corner).

[This page intentionally left blank]

ACTION PLAN

GENERAL INFORMATION

When you received your Disaster Management Plan, you received the template plan. When you have completed the assignments and added the required information to the printed text, you will have created a Disaster Management Plan customized for your company.

Do not leave blank any areas requesting information in your final plan. Type N/A (not applicable), DNA (Does Not Apply), or some other notation to alert the reader that the text area has not been ignored.

The printed output from text-software files is identical to the original printed text. See the DMP005-General Instructions file for printing specifications.

The software contains several files: text, spreadsheet and database.

1.0 NAMING CONVENTIONS

Section: Either a numeric (1.0) or alpha numeric. Example: "400 Action Plan designation.

Segment: A logical project management cycle for tasks requiring planning, delegation and assembly.

Phase: A logical grouping of sections and tasks to be performed.

2.0 ASSEMBLY INSTRUCTIONS

Your work will involve three phases:

1. Gathering data from existing materials within your company.
2. Researching new information using various sources.
3. Making decisions affecting organization structure, policy and procedure implementation, and related financial matters.

3.0 MANUAL SEGMENTS

The Disaster Management Planning Manual is divided into three segments, and instructions are provided for completing each section and references each file:

1. **Planning Segment:** The Planning Segment contains several devices, instructions and guidelines for gathering data.

   Please carefully review this material before performing any other action.

2. **Delegation Segment:** The Delegation Segment contains several devices for you to review carefully before assigning any task. The template memos in this

section require only your entry of names and dates. Section and page numbers are located in the lower left-hand corner of each page, preceded by the characters DMP.

3. **The Disaster Management Plan Segment**: The result of your research and development efforts. Section and page numbers are located in the lower left-hand corner of each page, preceded by the characters DMP.

4.0 DISASTER MANAGEMENT PLAN

The final **Disaster Management Plan** is divided into twenty seven sections.

To simplify your delegation and tracking, these sections should be completed according to the directions in the **ACTION PLAN**. Before you go any further; make three copies of the entire original printed manual, and store the original text.

1. One copy is for you to use as your note-taking device.
2. One copy is for you to use in assigning tasks as indicated in the Action Plan.
3. One copy is used to type the final plan on.

Create a 3 ring binder for the Master Plan with:

- Twenty seven numbered tabs: one for each section.
- Descriptive tabs for each office and department defined in Sections 13.0 - 16.0; Disaster Management Planning Manual and one tab for Section 2.0 The Introduction and Purpose (Disaster Management Planning Manual).
- Consider creating another tab in your copy only for **NOTES**, to accommodate those "things to do" that you want to accomplish at a later date.

5.0 STORAGE AREA

Create a storage area to contain all documentation, notes, drafts and other research material used to assemble the final plan. When the plan is complete, seal and store these materials.

Your company's legal advisor may have to defend the company's or your interests in court. If a court action is ever filed, you will be required to demonstrate why you created the plan as you did. Your notes and other research materials may be critical to your legal advisor's defense strategy.

Because the plan contains legal language, submit appropriate sections to your legal advisor if you do not clearly understand the potential ramifications of any part of this plan.

A special file named, Text Entry Template is designed for repetitive text entry of information required in several sections of the final plan. Because the plan is an evolutionary document, you must periodically change some of the information. Do not discard out - dated information: store it. You must provide the

Chairpersons and the Board of Directors with a summary of changes for your annual review, and using the out-dated information as a reference source may be necessary.

When you have completed the plan:

1. Create another binder with tabbed dividers only (no text), an exact duplicate of the binder format of the final Disaster Management Plan. Title it: UPDATED SECTIONS. As you replace out-dated sections or scripts, note the replaced sections as "Replaced: (Date)". Use this binder to store all changes.
2 Note the changes on pages provided in Section 18.0 of the Disaster Management Planning Manual (300 Disaster Recovery Plan) and store the revised pages in the UPDATED SECTIONS binder. Annually purge the changed information, and seal and store these materials.

[This page intentionally left blank]

PHASE 1 - PHILOSOPHY AND STRUCTURE

The entire plan is based upon the philosophy and structure contained in these sections. Review audit questions 1 and 3 in 400 Action Plan, PHASE 14 Examination and Audit Guidelines, for additional information.

Items for this phase are:

1. Policy and Procedure Statements.
   300 Disaster Management Plan, Section 1.0.
2. Introduction and Purpose of the Disaster Management Plan.
   300 Disaster Management Plan, Section 2.0.
3. Plan Development and Implementation - Guidelines and Scope.
   300 Disaster Management Plan, Section 3.0.

Tasks:

- For simplicity, the Disaster Management Planning Manual combines both policies and procedures into each section.
- Policies and commands are preceded by the terms "shall", "must", or "will", and procedures and suggestions are preceded by the terms "may" or "should".
- This may reflect regulatory guidelines for some industries.
- Review each provision, point-by-point, with your particular regulatory guidelines before adding or deleting any provision.
- Edit as appropriate, and make changes to reflect your company's names for functions and departments.

PHASE 2 - ARRANGE FOR SELECTION AND ASSEMBLY OF COMMITTEES

Arrange the selection and assembly of the Disaster Management Executive Committee; the appointment of the Disaster Management Team Chairpersons and Coordinators; and the selection of the Disaster Management Team members. Review items 2 and 9 in 400 Action Plan, PHASE 14 Examination & Audit Guidelines for additional information.

Items required for PHASE 2 are:

1. Meeting memo to be sent to Disaster Management Team members. DMP420 Samples (Memos), Exhibit 1, “Disaster Management and Business Recovery Plan”.
2. 100 Introduction, 1.0 Introduction to Disaster Management and Business Recovery. See also 400 Examination and Audit Guidelines.
3. 300 Disaster Response Guidelines, Section 8.3.
4. Disaster Management Team.
   300 Disaster Management Plan, Section 4.0.

5. Notification of Emergency Responsibilities.
   300 Disaster Management Plan, Section 5.0
6. Disaster Management Team: Duties and Responsibilities
   300 Disaster Management Plan, Section 6.1
7. Disaster Management Team Chairpersons: Duties and Responsibilities;
   300 Disaster Management Plan, Section 6.2;
8. Disaster Management Team Coordinators: Duties and Responsibilities;
   300 Disaster Management Plan, Section 6.3
9. Department Leaders: Duties and Responsibilities
   300 Disaster Management Plan, Section 6.4
10. Staff Personnel: Duties and Responsibilities
    300 Disaster Management Plan, Section 6.5 - 6.7.
    - Assistant Office Leader
    - Customer Service Representative
    - Office and Department Leaders

Tasks:

1. Schedule the initial meeting with all members of the Disaster Management Executive Committee. This initial meeting is preparatory to a combined meeting with all Disaster Management Team members.
2. One week before this meeting, send Section 2.0, Introduction and Purpose of the Disaster Management Plan; 300 Disaster Management Plan and Sections 4.0 - 10.0 of 300 Disaster Management Plan to all Disaster Management Executive Committee members, for review and discussion during the meeting.
3. Make necessary decisions during this meeting, for discussion during the combined meeting.
4. Review, edit and approve 300 Disaster Management Plan, Sections 4.0 - 10.0, entering all required Information.
5. Schedule the initial combined meeting of the Disaster Management Team and Executive Committee. This initial meeting is preparatory to assigning tasks to specific individuals. Prepare a copy of all sections included in this phase involving duties and responsibilities, for the Disaster Management Team to review and make recommendations upon.
6. One week before the combined meeting, send Section 6.1 - Section 6.7, 300 Disaster Management Plan, and related staff descriptions to each Disaster Management Team member (Office and Department Leaders) for review and comment during the meeting.

   Use the memo 420 Exhibit 1, Disaster Management & Business Recovery Plan for this task.
7. This may reflect regulatory guidelines for some industries. This will require the assembly of the Disaster Management Executive Committee, and the

approval of the duties and responsibilities of the Disaster Management Team, Chairpersons, Coordinators and Department Leaders.

8. Develop a Disaster Management Team Organization Chart and include the chart with this section, if appropriate. Use your company's routine organizational chart as a model. Decide who will fill these positions. Each position must have a dedicated backup. Edit and repeat categories as appropriate, and include any necessary name and title changes to reflect your company's functions and departments.
9. Consider using some of your Board members to fill certain positions, if appropriate, enter the required information.

Sample Format:

CO-CHAIRPERSON: [------- *Enter Name* -------], President/CEO

Responsible Board of Directors for:
Public and Media Relations
Shareholder and Stockholder Relations
Administration
Credit Administration
Backup: [... *NAME* ...]

CO-CHAIRPERSON: *[---- Enter Name ----],* [EVP, SVP, VP /ADMINISTRATION]

Responsible Board of Directors for:

Public and Media Relations
Administration
Personnel
Insurance
Accounting
Data Processing, Administrative Functions

Backup: [*Enter Name*]

CO - COORDINATOR: [*Enter Name*], [SVP, VP/OPERATIONS]

Responsible for:

Operations
Security
Telecommunications
Data Processing, Operations Functions
Backup: [*Enter Name*]

CO - COORDINATOR: [*Enter Name*], [VP/OPERATIONS, OPS SPECIALIST]

Responsible for:

- Operations
- Courier Services/Transportation
- Telecommunications
- Office Support
- Backup: [*Enter Name*]

MEMBER: [*Name*], [EVP, SVP, VP/CREDIT ADMINISTRATION]

Responsible for:

- Credit Administration
- Courier Services/Transportation
- Office Support
- Purchasing, Delivery and Supplies
- Backup: [*NAME*]

MEMBER: [*Enter Name*], [SVP, VP/ACCOUNTING]

Responsible for:

- Accounting
- Bookkeeping

Backup: [*Enter Name]*

MEMBER: [*Enter Name*], [VP, AUDITOR, COMPLIANCE MANAGER]

Responsible for:

- Audit
- Compliance, Legal
- Backup: [*NAME*]

MEMBERS: [*Enter Name*], [OPS SPECIALIST, COURIER, VENDOR]

Responsible for:

- Facilities
- Equipment and Supplies
- Transportation
- Backup: [*Enter Name*]

MEMBER: [*Enter Name*], [OFFICE MANAGER]

Responsible for:

- Office Operations
- Backup: [*NAME*]

MEMBER: [*NAME*], [ADM ASSISTANT, CLERK, CSR REP]

Responsible for:

- Disaster Records
- Backup: [*NAME*]

DMP300 Disaster Management Plan Sections 6.1 – 6.7 and related staff descriptions.

These sections address assigning special duties and responsibilities beyond the scope of the employee's position description.

Create a similar listing for each position within the company, or a common one for several positions having similar functions.

Enter the required information.

Place copies of routine position descriptions in Section 26.0 – Confidential Information.

PHASE 3 - REVIEW EMERGENCY NOTIFICATION PROCEDURES

Review Audit Questions; 7, 8, 10, 11, 12 and 13 in 400 Action Plan Phase 14, Examination & Audit Guidelines, for additional information.

Items required for this phase are:

- **Emergency Notification Procedures**
  300 Disaster Management Plan, Section 7.0
- **Emergency Response Guidelines**
  300 Disaster Management Plan, Section 8.0
- **Sample Forms**
  420 Samples - Forms, Chronological Log of Events
- **Sample Service Agreements:**
  420 Samples - Sample Agreements
- **Service Agreements, Emergency Services Agencies and Community Resources.**
  420 Disaster Recovery Plan, Agreements & Emergency Services.
- **Sample database screens.**

420 Disaster Recovery Plan, Samples - Memos & Data Base screen templates.

- **Centers of Operations.**
  300 Disaster Recovery Plan, Section 10.0.

Tasks:

- This section addresses the logical progression of events before, during and after an emergency. This section also describes the plan's documentation mechanism.
- Review and edit/approve 300 Disaster Management Plan, and 420 Samples - Forms, "Disaster Recovery Log".
- Choose appropriate staff personnel to research, develop and input the requested information.
- This is a suggested list only, based upon the assumption that a phone book may be unavailable during the disaster.
- Consider adding your standard vendor information to this list, and creating a database file to contain all entries.
- Print and distribute this list on a quarterly/semi - annual/annual basis.
- Use 420 Sample - Memos, Data Base Screens; as a guide.
- This may reflect regulatory guidelines for some industries.
- Review each provision, point - by - point, with your particular regulatory guidelines before adding or deleting any provision.
- This section addresses assigning alternate emergency operations sites, and may be edited to reflect sites for storage, data processing, cash-handling and business facilities.
- Decide which alternate sites to use, and enter the required information.
- 300 Disaster Management Plan, Section 10.0, Centers of Operation
  - Primary Center of Operations.
  - Alternate operation sites # 1 - # 4;.
  - Storage site.

PHASE 4 - RISK ASSESSMENT ANALYSIS

Review Audit Questions: 4, 5, 6, 7, 8, 11, and 13 found in 400 Action Plan, PHASE 14 Examination and Audit Guideline Questions, for additional information.

Items required for this phase are:

1.0 300 Disaster Management Plan, Section 12.0, Risk Assessment and Evaluation.

2.0 420 Samples - Memos, Exhibit 12, Disaster Management Recovery Questionnaire memo.

3.0 420 Sample- Facility diagrams.

4.0 300 Disaster Management Plan, Section 13.0, Department Recovery Plan Administration.

5.0 300 Disaster Management Plan, Section 14.0, Department Recovery Plan, Cash handling Facility.

6.0 300 Disaster Management Plan, Section 15.0, Department Recovery Plan, Data Processing.

7.0 300 Disaster Management Plan, Section 16.0, Department Recovery Plan, Generic.

8.0 420 Sample memos, Exhibit 14, Emergency Supplies and Equipment.

Tasks:

1.0 Schedule a second meeting of the Disaster Management Team within two weeks of the initial meeting.

2.0 One week before the combined meeting, send Sections 12.0, 300 Disaster Management Plan, Annual Risk Assessment and Evaluation to all Disaster Management Team members for review and discussion during the meeting.

3.0 Use the memo 420 – Sample Memos, "Disaster Management & Business Recovery Plan Orientation" for this task.

4.0 During this meeting, assign the completion of the first draft of Sections 6.0 - 6.7 to Department Leaders.

300 Disaster Management Plan, Section 13.0, Department Recovery Plan Administration.

300 Disaster Management Plan, Section 14.0, Department Recovery Plan – Cash Facility Recovery Plan;

300 Disaster Management Plan, Section 15.0, Department Recovery Plan for Data Processing

300 Disaster Management Plan, Section 16.0 Department Recovery Plan - general is the template for **non** - cash-handling facilities where there is no other specific template.

5.0 This may reflect regulatory guidelines for some industries.

6.0 Prioritize this list according to your geographical area and re-order if necessary.

7.0 Enter the required information.

8.0 Create an Emergency Exit diagram for each facility.

Include an appropriate exit diagram with each Department Recover Plan, Sections 13.0, 14.0, 15.0 and 16.0 in 300 Disaster Management Plan.

This may reflect regulatory guidelines for some industries.

9.0 The accurate completion of the individual recovery scripts is the most important assignment of the plan. It may be necessary to expand or delete information, depending upon the needs of your company.

10.0 The Department Leaders should enter preliminary information in Draft #1 and return the drafts to you for review.

11.0 The information requested in the templates is the suggested minimum information to be developed.

12.0 Topics for special discussion are listed below, and the topics follow the master format for all Department Recovery Plans, Sections 13.0 – 16.0

13.0 Complete these suggestions as you feel they apply to your company.

14.0 Sample Format for Department Recovery Plans.

14.1 Consider dedicating a team of employees to handle special transportation and delivery tasks.

14.2 In the event a disaster disrupts telephone communications, it is likely that only the trunk (area code) line will be affected. To plan for the necessary redundancy in communications ability, your Bell system (not other private systems) will likely restore services according to this scenario:

Phone company internal communications.

- Phone company overall network.
- Area code ladder.
- Public phones (Bell system only).
- National security links.

Essential services (health and welfare agencies).

- American Red Cross.
- Emergency Operations Centers (state and county).
- Police/Fire/Paramedic/Ambulance.
- Companies (call your local Office of Emergency Services to inquire about priority re-installation).

- Food services.
- General customers:
  - Commercial.
  - Residential.
  - Lifeline.

Normal; and Other systems (Private vendors: private internal systems do not link directly to Bell systems, and likely will not receive priority re-installation).

Cellular telephone links depend upon the location of repeaters, and not upon a Bell system.

14.3 Public Phone # 1: 555-555-5555

Location: N/W wall of Thrifty Drug, located at 2900 Belvedere Dr. at front door

14.4 List appropriate numbers for employee hotline and security alarm company:

- Dedicated company phone line.
- Off-site dedicated line.
- Answering service to handle employee communications only.

14.5 Include the department's critical functions, in order of priority.

14.6 Include the department's accessory functions, in order of priority.

14.7 **PRIMARY:** e.g., Northwest corner of Thrifty Drug parking lot, located at 2900 Belvedere Dr.

SECONDARY: e.g., Southwest corner of building's parking lot, located at the intersection of Third Ave. and Main St.

Develop diagrams for all facilities. Direct the location and storage of accurate blueprints of each facility, and develop a scaled representation of each facility to include the: Location of the emergency staging areas;

Location of all exits, with direction of swing for all doors and windows; and the Location of all utility shut-offs (color- coded):

WATER:	Blue
GAS:	Green
ELECTRICAL:	Red

If you feel it appropriate, direct the painting of the utility shut-off devices the same coded colors.

14.8 **PRIMARY:** e.g., California Street Office, located at 5573 California Street, Your City, Your State; 99999, 555-555-1212 FAX: 555-555-1213

SECONDARY: e.g., Any Office Building, located at 1275 Fourth Street, Your City, Your State 99999; 555-555-1214 FAX: 555-555-1215

14.9 Create an actual list of necessary records and equipment.

14.10 List special personnel, customer, vendor, facility, asset and records needs.

14.11 **CASH**: For cash-handling facilities, include a vault or cash carts. If another company is to be used, the other company should supply all appropriate maps, diagrams and instructions for use of its facility. A complete office operations manual department should be stored at this site.

14.12 **DP**: For data processing, include air conditioning requirements. If a hot or cold site vendor is to be used, the vendor should supply all appropriate maps, diagrams and instructions for use of its facility.

14.13 If a hot or cold site vendor is to be used, the vendor should supply written operational procedures, fully compatible with the company's plan. A complete operations manual for the company's data processing department should be stored at the hot or cold sites.

14.14 Current written operational procedures for offices, all departments and functions should be stored at each of the centers of operations, if available.

14.15 Office supplies, emergency equipment and survival supplies should be assembled and stored within the each facility, as well as emergency medical supplies and an appropriate supply of critical forms.

14.16 List and describe appropriate forms; and use DMP420 Sample memo, Emergency Supplies for Office", as a guide for your supplies and equipment acquisition.

PHASE 5 - TESTING AND TRAINING REQUIREMENTS

Review audit questions 14 and 18 in DMP400 Action Plan Audit Questions for additional information.

1.0 DMP300 Disaster Management Plan, Section 11.0 Testing and Training Requirements.

2.0 DMP300 Disaster Management Plan, Section 17.0 Training Program.

3.0 DMP300 Disaster Management Plan, Section 27.0 Leaders Guide.

Tasks:

1.0 Decide upon the frequency and method of testing.

2.0 Coordinate Disaster Management Plan training with other security training.

2.1 Sections 13.0 – 16.0 of the Disaster Recovery Plan must be completed before completing the Leader's Guide and Training Handout.

2.2 Decide upon the personnel categories to be trained (staff, supervisors, managers, executives, Board members).

2.3 Edit/approve the information and forms contained in both the Leader's Guide and Training Handout. Because both training documents contain statements regarding philosophy, policy and procedure, review each one carefully, and make changes where appropriate.

2.4 Enter the required information in both the Leader's Guide and the Training Handout and add attachments as required.

3.0 420 Sample Memos, Agreements, Forms and Handouts - Training Handout

3.1 Consider adding appropriate information from various handbooks, Instructional materials and other company manuals.

3.2 Use this handout for the company's annual training program, either as a structured delivery or a "read and initial" device.

4.0 300 Disaster Management Plan, Section 27.0 – LEADERS TRAINING GUIDE

4.1 Consider adding appropriate information from various handbooks, instructional materials and other company manuals.

4.2 Use this guide for the company's annual training program.

PHASE 6 - REVIEW ALL RECOVERY SCRIPTS

Review audit questions: 5, 6, 7, 11, and 13 in 400 ACTION PLAN, PHASE 14, Examination and Audit Guideline Questions, for additional information.

Items required for this phase:

All department recovery scripts, (Draft #1).

400 Sample Memos, Exhibit 3, "Annual Test and Training Program" memo.

400 Sample Memos, Exhibit 9, "Equipment Data Base fields", memo.

400 Sample Memos, Exhibit 14, "Emergency supplies for Office and Departments" memo.

Tasks:

1.0 Collect Department Recovery Plans, Sections 13.0 – 16.0, from all Office and Department Leaders.

2.0 Based upon the draft results, either meet with appropriate Disaster Management Team members for review and re-draft, or send the sections back for edit.

3.0 Use the memo in 400 Sample Memos, Exhibit 2, "Annual Testing and Training Program" for this task.

4.0 Analyze the information contained all Department Recovery Plan scripts (Sections 13.0 - 16.0).

5.0 Develop a checklist of items to be acquired or completed, including projected costs.

6.0 Use the checklist contained in 420 Sample Memos, Exhibit 14, "Emergency Supplies for Office and Departments", for this task.

7.0 Rather than order additional office and related supplies already in stock, simply ensure that all locations maintain at least a five (5) -day inventory of these items.

8.0 Create a tentative budget, based upon the information contained in the completed Department Recovery Plans.

9.0 Enter all required information in each Department Recovery Plan, sections 13.0 – 16.0.

PHASE 7 - REVIEW ALL RESOLUTIONS FOR THE DISASTER MANAGEMENT PLAN

Review audit questions 2, 9, and 16 in 400 Action Plan, PHASE 14: Examination and Audit Guide Line Questions for additional information, and assemble the Disaster Management Executive Committee. Items required for this phase are:

- Resolutions for the Disaster Management Plan; see 300 Disaster Management Plan, section 20.0.
- Disaster Management Executive Committee: Appointment; ref. 300 Disaster Management Plan, Section 21.0.
- Disaster Management Executive Committee: Acknowledgment; ref. 300 Disaster Management Plan, Section 22.0.
- Board of Directors' Certification Resolution, ref. 300 Disaster Management Plan, section 23.0.
- Board of Directors' Re-certification Resolution, ref. 300 Disaster Management Plan section 24.0.

Tasks:

1.0 This Phase may reflect regulatory guidelines for some industries.

2.0 The Board of Directors' resolution authorizing the members of the company's Disaster Management Executive Committee and Team to assume and perform extraordinary duties and responsibilities under emergency circumstances.

3.0 The resolution insures the continuity of management and business operations in the event of a declared emergency.

4.0 Add line of succession names.

5.0 Add Center of Operations information.

6.0 Add alternate site information.

7.0 The Board of Directors' resolution appointing the members of the company's Disaster Management Executive Committee.

8.0 Add names and titles as required;

9.0 Obtain Board Secretary's signature and date; and distribute a Board Secretary - signed copy to each appointee.

10.0 This may reflect regulatory guidelines for some industries.

The appointee's acknowledgment of membership on the company's Disaster Management Executive Committee.

11.0 Add names and titles as required.

Prepare one (1) copy and one (1) original for each appointee; and obtain appointee's signature and date, and return the signed copy to the appointee. (Ref. Phase 12.)

12.0 This may reflect regulatory guidelines for some industries.
Original certification of the Disaster Management Plan by the Board of Directors.

13.0 Obtain Board Secretary's signature and date after initial review. (Ref. Phase 12.)

14.0 This may reflect regulatory guidelines for some industries.

15.0 Annual re-certification of the Disaster Management Plan by the Board of Directors. Obtain Board Secretary's signature and date after annual review. Disaster Management Plan, Section 24.0.

16.0 Storage location of appropriate documents.

17.0 Arrange for appropriate storage of required items.

18.0 Complete all required information.

19.0 When the plan has been approved, place full copies of the plan at the storage locations.

PHASE 8 - REVIEW CONTRACTS / INSURANCE POLICIES / CORPORATE RECORDS

Review Audit Guideline questions: 4, 8, 15 and 17 in 400 Action Plan, "Examination and Audit Guidelines" for additional information. Items required for this phase are the original documents for:

- Contracts.
- Insurance policies.
- Appropriate corporate records.

Tasks:

1.0 This Phase may reflect regulatory guidelines for some industries.

2.0 Original records that are not also stored as electronic data should not be used in the workplace: use only photocopies of these documents.

3.0 Ensure that all the listed documents are consolidated at a central location, indexed, copied and stored.

4.0 Consider using a database or spreadsheet program for this task.

PHASE 9 - CENTERS OF OPERATIONS AND RECOVERY SCRIPTS

Review Audit Guidelines questions 5, 6, 7, 11, and 13 in 400 ACTION PLAN, Phase 14: "Examination and Audit Guide Line Questions" for additional information.

Items required for this phase:

- 400 Action Plan, PHASE 9.0, "Centers of Operations and Recovery Scripts".
- 300 Disaster Management Plan, Sections 13.0 – 16.0 Recovery scripts.
- 300 APPENDIX A – Sample Memos, Exhibit 14, "Emergency Supplies and equipment for Office and Departments" memo.

Tasks:

Develop final budget and have your budget approved for the replacement or acquisition of all necessary items, and the use of any contract facilities:

- Facilities (all).
- Office supplies.
- Equipment.
- Vendors.
- First aid kits.
- Food and water supplies.
- Fire extinguishers.
- Operations manuals.
- Printing and copying.
- Keys and alarm access codes.
- Disabled access.
- Power supplies.
- Conventional.
- Alternative.
- Blueprints, diagrams and maps, if appropriate.
- Delivery plan for personnel, equipment and supplies.
- Cash-handling facilities (special needs):
    - Currency and coin.
    - Negotiable documents.

- Acquire and distribute all emergency supplies and equipment to appropriate offices, departments and facilities.
- Have the Office or Department Leader personally sign an inventory receipt for these items.
- Forward a copy of the signed inventory list to the Auditor.

PHASE 10 - FORMS

Review Audit Question18, DMP400 Action Plan, for additional information.

Items required for this phase:

- 300 Disaster Management Plan, Section 19.0 Sample Forms.
- 300 Disaster Management Plan, Sample forms, Exhibit 3, "Testing of Solutions Documentation".
- 420 Sample Forms, Exhibit 4, "Recovery Worksheet".
- 420 Sample Forms, Exhibit 8, "Employee Personal Profile".

Tasks:

1.0 300 Disaster Management Plan, Section 15.0 Department Recovery Plan - Data Processing Review, edit and approve. If you feel that other forms should be added, or these forms deleted, enter that information.

2.0 300 Disaster Management Plan Section 19, Sample Forms, Exhibit 2, "Chronological Log of Events", Review, edit and approve. This form should be completed by the appropriate Office or Department Leaders after any test of the plan.

3.0 300 Disaster Management Plan Section 19, Sample Forms Exhibit 3, "Testing of Solutions Documentation". Review, edit and approve. A copy of this Worksheet should be included with every copy of the final plan.

4.0 300 Disaster Management Plan, Sample Forms, Exhibit 4, Recovery Work Sheet for:______________ Distribute this form, and provide an envelope for the employee to return the form to you or your designate.

5.0 Distribute a copy of the Employee Personal Profile form to each employee.

6.0 Request the Chairpersons distribute a copy to each Board Member.

7.0 Store all returned forms in a secure location.

PHASE 11 - CONFIDENTIAL PERSONNEL INFORMATION

Items for this phase:

300 Disaster Management Plan, Section 26.0, Confidential Information, and Personnel database printouts.

Tasks:

Whatever lists were not placed in Section 18.0, Update Information, place in 300 Disaster Management Plan, Section 26.0, Confidential Information

PHASE 12 - POLICY AND PROCEDURE STATEMENTS

Review Audit Guideline Questions 2, 14, 16, and 19 in 400 – Examination and Audit designation for additional information.

Items required for this phase:

- 300 Disaster Management Plan, Section 1.0 Policy and Procedure Statements.
- 300 Disaster Management Plan, Section 6.1 Disaster Management Team: Duties and Responsibilities.
- 300 Disaster Management Plan, Section 10.0 Centers of Operations.
- 300 Disaster Management Plan, Section 18.0 Update Information.
- 300 Disaster Management Plan, Section 20.0 Resolutions for the Disaster Management Plan.
- 300 Disaster Management Plan, Section 21.0 Disaster Management Executive Committee Appointment.
- 300 Disaster Management Plan, Section 23.0 Board of Directors Certification.
- 300 Disaster Management Plan, Section 27.0 Leaders Training Guide.
- 420 Sample Memos, Exhibit 5, “Synopsis Disaster Management Plan.

Tasks:

1.0 The final plan must be presented to the company's Board of Directors for initial approval, and annually from then on. For both the initial and annual reviews, perform the following tasks.

2.0 Have the Chairpersons schedule the Board presentation.

3.0 Prepare two (2) binders for each Board member, to be delivered during the scheduled meeting, containing:

- 300 Sample Memos, Exhibit 4, “Memo to Board of Directors Business Recovery”.
- 300 Training Handout.

- 300 Sample Memos, Exhibit 5, "Synopsis Disaster Management Plan" and current copies of Sections 4.0, Disaster Management Team and Section 9.0, Service Agreements, Emergency Services Agencies and Community Resources.

4.0 Prepare one complete plan for individual member review, if desired, for general Board review before the scheduled meeting.

5.0 Prepare copies of documents for the Board Secretary's signature (new documents if this is the initial review, revised documents if this is an annual review):

- 300 Disaster Management Plan, Section 20.0; Resolutions for the Disaster Management Plan.
- 300 Disaster Management Plan, Section 21.0; Disaster Management Executive Committee: Appointment.
- 300 Disaster Management Plan, Section 22.0, Disaster Management Executive Committee: Acknowledgement.
- 300 Disaster Management Plan, Section 23, Board of Directors Re-Certification Resolution.

6.0 Initially and annually, present a synopsis of the Disaster Management Plan to the Board of Directors for review and approval, or redesign if necessary.

Use 420 Sample Memos, Exhibit 5, "Synopsis, Disaster Management Plan" for this task; and Use Section 22.0, Update Information as an updating device.

The annual review should contain your certification that the plan has been tested, and revised where appropriate.

Use or edit Sample Memo for this task.

Place a copy of the signed resolution(s), or a copy of the Board minutes indicating approval, in the master copy of the plan for periodic review by regulatory agencies.

Distribute copies of the plan and additional copies of the forms contained in Section 19.0, 300 Disaster Management Plan, according to the guidelines in Section 1.4, Distribution of Statements.

PHASE 13 - UPDATE INFORMATION

Review Audit Guideline questions 4, 15, 18, and the HYPOTHETICAL DISASTERS section in 400 Action Plan, for additional information.

Items required for this phase:

300 Disaster Management Plan, Section 18.0 and 19.0 Update Information and Forms.

420 Sample Forms, Exhibit 3," Testing of Solutions Documentation".

300 Disaster Management Plan, Section 27.0, "Leader's Training Guide" and required attachments.

420 Sample Memos, Training Handout.

420 Sample Memos, Exhibit 2, "Annual Risk Assessment Exercise".

420 Sample Memos, Exhibit 13, "Annual Plan Review".

Tasks:

1.0 Select one of the problem-solving exercises from 400 Action Plan, Phase 14, to be used for the annual test for every office, department and facility.

2.0 Keep a copy of each test in a binder marked TESTS for auditing and examination purposes.

Have Office and Department Leaders schedule and conduct the annual risk assessment, review and training assignments:

- 400 Action Plan, Phase 4, Risk Assessment Analysis.
- 300 Disaster Management Plan, Section 12.0, Annual Risk Assessment and Evaluation
- 400 Action Plan, Phase 5, Annual Testing and Training Requirements.
- Address necessary revisions and note changes in 300 Disaster Management Plan, Section 18, Update Information.
- Document training as indicated in 300 Disaster Management Plan, Department Leaders, section 6.4.
- Document inspection and testing as indicated in 420 Sample Forms, Exhibit 3, Testing of Solutions.
- For all departments except Data Processing, a written critique as a group exercise is acceptable.

3.0 Have the Data Processing Leader conduct independent physical tests as scheduled, and annually provide you with completed 420 Sample Forms, Exhibit 3, "Testing of Solutions", for each backup facility tested.

4.0 For the Data Processing Department, only a physical test of the backup plan should be considered acceptable.

5.0 Calendar and prepare for future.

6.0 Auditing and updating of policies and procedures.

7.0 Annual risk assessment exercise.

8.0 Physical inspection of facilities and supplies.

9.0 Testing of the plan by each office and department.

10.0 At least Annual training.

11.0 At least an annual updating of all database lists.

12.0 At least an annual updating of all copies of the Disaster Management Plan; and an annual review and certification by the Board of Directors.

PHASE 14 - EXAMINATION AND AUDIT GUIDELINE QUESTIONS

A business recovery plan has many descriptive names: disaster management plan, disaster recovery plan, emergency preparedness plan, contingency plan, business resumption plan, disaster contingency plan and so on. For simplicity, the term "contingency plan" is used throughout this material.

Developing a contingency plan for a company is one thing: developing a plan that is effective and that auditors and examiners will support and approve is another. A contingency plan should not be a secret or a mystery, but just a routine part of business planning and operations.

An effective contingency plan simply covers all the bases of the company's business operations. This means personnel, customers, facilities, functions, assets and records; in brief, everything.

This means a straightforward, uncomplicated approach to resuming business operations in the event of a disaster.

An effective contingency plan is developed as if all the key players will be absent. This means written procedures that support policies; a clean format that is easy to understand and implement; business tools with which to work; and most important of all, full management support and guidance from start to finish.

On the following pages is a suggested examination checklist for the entire organization, not data processing only. Additional guidelines and suggestions are available in other publications available in the private sector.

These are basic questions to be answered during an audit or examination. If the company has not adequately addressed these issues, it may be unable to restore even basic services after a disaster. This worksheet contains both questions and the rationale or explanation for an examiner's inquiry.

PHASE 14 - Examination and Audit Guideline Questions

Audit Questions	YES	NO
1. **Has the company received a commitment from its Board of Directors and senior management to develop a contingency plan, on a company-wide basis?** This should be a written directive from the company's Board of Directors or the President/CEO. This directive should clearly state the company's intention to implement a plan, according to the guidelines issued by local, state federal or any other regulatory agency, and contain time-lines for completion.		
2. **Established a management group responsible for the development and Implementation of a contingency plan?** This management group should consist of executives and senior department managers, and Board members if appropriate. If a data processing service bureau or other critical vendor is involved, a senior member of the service bureau or vendor company should also be a member of this group. Other vendors should be represented as appropriate. The written directive should contain this information. Those persons responsible for the succession of management and recovery efforts should be designated by name in the contingency plan: • Line of succession for the Board of Directors. • Line of succession for the company's management. • President/CEO: Leadership and media/public relations at the corporate level, and notification to and interaction with any regulatory agency of a disaster. • Chairpersons (2): Leadership, media/public relations and executive decisions at the company level. • Coordinators (2): Implementation, media/public relations and follow-through at all levels; and Office/department/functional leaders (2): Leadership, implementation and follow-through at office, department or functional level.		

EXAMINATION AND AUDIT GUIDELINE QUESTIONS

Audit Questions	YES	NO
3. Defined the components of, and structural framework and appropriate methods for creating and updating the company's contingency plan? No two companies are structurally alike, and a contingency plan should be developed in consideration of the organization's actual needs. A sensible contingency plan contains at least three (3) core elements: • Management business resumption plans. • Emergency procedures. • Information systems' data processing recovery procedures. Considering the massive amount of information and data requiring input, updating and storage, the company should assemble its contingency plan in two (2) phases: • **Text:** All information requiring only original input and printout, minimal updating and distribution, approval, and no sorting. • **Database**: All information requiring continual changes of data affecting the entire plan, frequent updating and distribution, no approval, and all lists requiring sorting. The contingency plan should contain a table of contents, and an appendix if appropriate. Copies of the approved contingency plan should be distributed to each office, department and facility designated as an alternate operations site. A mechanism should be developed to update contingency plan information at least annually.		
4. Conducted and documented a risk assessment analysis by the management group? It is impractical to create a contingency plan to address unlikely potential risks. The risk assessment analysis should concentrate on real, likely or historical risks, and should include: • A definition of a disaster as any event that will significantly impact the company's operations; and • Potential threats by natural, technological and human causes. Prioritized disasters likely to occur, based upon a historical review of the region's or the company's: • Weather and geographical conditions (fire, flood, storm, earthquake, adjacent hazardous businesses);		

EXAMINATION AND AUDIT GUIDELINE QUESTIONS

Audit Questions	YES	NO
• Facilities' issues (age, condition, repairs). Equipment issues (age, condition, repairs, hardware/software failure, compatibility). • Utilities' services (interruption of communications, water, electricity, gas, sewer); and Human issues (robberies, strikes, riots, sabotage, bomb threats). An assessment of the potential impact resulting from loss of information and services to: • Financial condition. • Competitive position. • Customer confidence. • Legal/insurance/regulatory requirements. An analysis of the anticipated short-term and long-term costs to minimize exposure, including appropriate insurance coverage for: • Directors and officers. • Casualty claims, both employee and customer. • Property damage, both company and vendor owned. • Business interruption costs.		
5. Evaluated, prioritized and developed written critical and accessory functions for the continuing operation of all departments, offices, facilities, functions, and personnel? Every function contains critical tasks that must be performed, and accessory tasks that may be performed after critical tasks have been completed. An appropriate response to disaster recovery dedicates all available resources to stabilizing critical functions first, and then restoring less important functions.		

EXAMINATION AND AUDIT GUIDELINE QUESTIONS

Audit Questions	YES	NO
6. Developed, maintained and updated comprehensive written policies and procedures governing the routine duties and responsibilities of all departments, offices, functions and personnel? Written in the form of an operations manual, or "desktop procedures", these guides provide immediate help to persons involved in both day-to-day and exceptional disaster recovery tasks. Cross-training employees to perform tasks in more than one role increases the likelihood of a successful recovery from the disaster. Lack of a comprehensive policy manual may indicate ineffective leadership and direction from management. This may also promote arbitrary and capricious decisions by personnel at all levels within the company, based upon an individual's interpretation of a policy that is "understood", but not committed to writing. Lack of a comprehensive operations and procedural manual may indicate ineffective planning and preparation for the accomplishment of tasks. This may also promote arbitrary and capricious decisions by personnel at all levels within the company, based upon an individual's interpretation of a procedure that is commonly accepted, but not committed to writing. Functional responsibilities for both departments and offices that are informal and unwritten, poorly defined or nonexistent, may lead to confusion, ineffective prioritization of tasks, and misinterpretation of directions. This also makes assessing the responsibility for results difficult, if not impossible.		

EXAMINATION AND AUDIT GUIDELINE QUESTIONS

Audit Questions	YES	NO
7. Developed, maintained and documented written strategic recovery procedures for all departments, offices, facilities, functions, and personnel? Strategic recovery procedures are designed to dedicate appropriately the company's resources to: • Minimize disruptions of services to the company and its customers. • Minimize financial loss. • Ensure a timely resumption of operations in the event of a disaster. • Create a strategic and tactical response based upon these priorities: ▪ People: Employees and customers. ▪ Places: Offices, departments and facilities. ▪ Things: Assets and records.		
8. Developed, maintained and distributed listings of individuals and companies who will help the company recover from the disaster? The organization should take advantage of all available resources for its disaster recovery efforts. Contacting these resources in a timely manner is critical. This component should include at least listings of appropriate: • Personnel (from Personnel Department); • Emergency services and local governmental agencies (from telephone book); • Vendors (from Form 1099 list). • Equipment (from fixed asset list). • Transportation, both public and private (from telephone book); • Media representatives (from telephone book and personal contact, include FAX numbers). • Insurance policies and representatives (from original policies). Additional information should be available for all facilities, to include diagrams demonstrating the location of: ▪ Emergency staging areas. ▪ Evacuation routes. ▪ Public telephones and numbers. ▪ Utility shut-off devices.		

EXAMINATION AND AUDIT GUIDELINE QUESTIONS

Audit Questions	YES	NO
9. Developed, maintained and periodically reviewed written position descriptions, levels of authority and reporting routines for all personnel, for both routine and exceptional operations? This should provide estimates for the minimum and maximum number of hours acceptable to restore basic functions, and should include at least the company's: • Administration (Board of Directors, executives.). • Acceptable maximum: (08-24 hours). • Operations (both total company and individual office operations; telecommunications; security; personnel. • customer service; purchasing, delivery and couriers. • Credit and Accounting functions, and office support. • Acceptable maximum: (24-48 hours). • Data Processing (both in-house departments and service bureaus.) Acceptable maximum:(24-48 hours).		
10. Evaluated the adequacy of contingency plans for its service bureau, if one is used, and ensured that the company's contingency plan is compatible with its service bureau's plan? This should include a written recovery plan created by the service bureau, similar in format to the office, department and function recovery plans created by the company, and furnish: • Names and phone numbers of the persons responsible for recovery. • Physical addresses of, phone and FAX numbers, and maps and directions to the facilities to be used for recovery efforts. • Appropriate access and security information; • Recovery steps and time-lines for each function to be restored; and Costs for services.		
11. Developed, acquired or contracted to use at least one alternate location to house critical functions on a temporary basis? If the company maintains several locations and facilities, shifting strategic operations to a designated alternate site is simplified. If the company does not maintain another location or facility, it should arrange to use a facility controlled by another similar company. This should include access to a(n) off-site facility(s) to: • Store copies of the company's contingency plan. • Allow the company to resume administrative functions. • Allow the company to resume operations functions.		

Audit Questions	YES	NO
• Allow the company to resume non-service bureau data processing functions, and store backup copies of daily data files, program libraries and related data processing supplies. The information supplied by the provider company should include: ▪ Names and phone numbers of the persons responsible for recovery. ▪ Physical addresses of, phone and FAX numbers, and maps and directions to the facilities to be used for recovery efforts. ▪ Appropriate access and security information. ▪ Recovery steps and time-lines for each function to be restored and Costs for services.		
12. Obtained written backup/recovery agreements to service appropriate critical functions? These operations often represent critical services to companies and their customers. The loss or extended disruption of these business operations poses substantial risk of financial loss. A written contract commands performance, and assures the company of appropriate access and service. Written agreements should include: Cash-handling facilities and appropriate personnel, equipment and supplies (customer service functions); Non-cash-handling facilities and appropriate personnel, equipment and supplies (administrative functions); and Data Processing: • Central computer and distributed processing. • End user computing. • Local area networking. • Nationwide telecommunications. • Correspondent companies and clearing houses (ACH).		
13. Stored appropriate emergency supplies of food, water, first aid kits, fire extinguishers and basic tools? hours or days, depending upon the magnitude of the disaster. They may require nourishment, sanitation facilities and light. Both employees and customers may be injured during the disaster, and medical care may not be immediately available. They may require first aid supplies and written instructions. The work location may suffer structural and utility damage, placing employees and customers at risk. Basic tools and other non-business supplies should be available to help employees deal with these hazards. Suggested items for placement in offices Both employees and customers may be trapped at a work location for, departments and facilities are: • Basic tools (hand axe with a hammerhead at one end, vice-grips, small pry-bar, small shovel, Swiss army-type knife, pipe wrench, channel lock pliers, 1" wide stiff putty knife, wire cutters, both blade and Phillips screwdrivers);		

Audit Questions	YES	NO
• Two (2) pair of heavy-duty leather work gloves (size large), and one (1) box of 100 count latex surgical gloves; • One (1) battery-powered AM/FM radio, and extra batteries; • Two (2) flashlights and twice the number of batteries necessary to power each one. One (1) package of light sticks; several candles and matches; • Three (3) 50' rolls of duct tape; one (1) 50' roll of plastic electrical tape; and two (2) 25' rolls of strapping tape; • One (1) 100' coil of 1/4" nylon rope, and one (1) roll of yellow emergency "CAUTION" or "WARNING" marking ribbon; • One (1) box each of 100 count "tall kitchen" and 33 gallon plastic heavy-duty garbage bags with ties; • "Moon" or metal-ized emergency blankets; and Sanitary napkins and other similar absorbent materials.		
14. Developed, maintained and delivered a written contingency plan training program to employees, at all levels within the company? Training all employees regarding their duties and responsibilities during disaster recovery efforts is critical to the contingency plan's success. The delivery of standardized training for all employees results in a disaster recovery teamwork effort, and promotes reasonable behavioral expectations. Employees who are knowledgeable about the company's procedures, the risks associated with each employee's role within the company, and appropriate disaster recovery techniques, reduce the potential for operational losses. Actual hands-on training eliminates developmental errors, improves procedures, and reduces the opportunity for mis-communication. Experiential training promotes employee comfort and confidence. An effective training program has three (3) components: • **Orientation**: All new employees, regardless of position description, should be oriented regarding the organization's philosophy, organizational and reporting structure, goals, priorities and other appropriate information. • **Transfer/Promotion**: Any employee who has been transferred to a new department, or has received a promotion, should be thoroughly trained regarding his/her new duties and responsibilities, and the policies, procedures and control mechanisms used in this new capacity. • **Updating**: Any change in management, business philosophy, technology, legal requirements, policy or procedure should require re-training of all involved employees. Periodic exposure increases effectiveness.		

<table>
<tr><th>Audit Questions</th><th>YES</th><th>NO</th></tr>
<tr><td>15. Developed, maintained and implemented an effective storage and recovery plan for the organization's original documents and vital records?

Recovering business operations after a disaster often requires the use of original documents and vital records not stored as electronic data. The contingency plan should include:

The consolidation and storage of appropriate original documents and vital records in a central fire-proofed location:
• Contracts.
• Insurance policies.
• Corporate papers.
• An inventory list of stored items, stored in two (2) locations.
• Annual review for applicability, currency and legality.</td><td></td><td></td></tr>
<tr><td>16. Developed, maintained and documented an effective annual reassessment of the company's contingency plan?

A contingency plan should be an evolutionary program, designed to adapt to changes as they occur.

This should be a synopsis of all changes made to the contingency plan since the last annual reassessment. This should be presented to the Board of Directors for review and approval. This annual approval should be reflected in the Board of Directors' minutes.</td><td></td><td></td></tr>
<tr><td>17. Developed, maintained and documented an effective annual reassessment of the company's insurance coverage?

This should be a synopsis of all changes made to the contingency plan requiring an accompanying change in insurance coverage, since the last annual reassessment. This insurance coverage reassessment should include all policies addressing:
• Directors and officers.
• Casualty claims, both employee and customer.
• Property damage, both company and vendor owned.
• Business interruption costs.

This should be presented to the Board of Directors for review and approval. This annual approval should be reflected in the Board of Directors' minutes.</td><td></td><td></td></tr>
</table>

Audit Questions	YES	NO
18. Developed, maintained and documented the criteria, conditions and frequency for testing the contingency plan, and has the company actually performed, evaluated and documented a test? Testing all aspects of the contingency plan at least annually is critical to the contingency plan's success. Actual hands-on testing eliminates developmental errors, improves procedures, and tests the interaction of the contingency plan's components. It is physically impractical to move an office, department or function to another location as a test. The use of a hypothetical problem-solving exercise, in a documented meeting environment, will provide the necessary test results. Physically testing the company's data processing backup capabilities is essential. An appropriate physical test should be conducted at least annually, at each of the company's designated recovery facilities.		
19. Have copies of the approved plan been distributed appropriately? Distribution should involve offices, departments, facilities and functions, and all alternate operations sites. After the contingency plan has been approved, copies of the plan and appropriate lists should be distributed. Each recipient should receive two (2) copies: one to keep at work, and one to keep at home or the truck of his/her primary vehicle. Full copies of the contingency plan should be delivered to the Chairpersons and Coordinators, and stored at each alternate site. Limited copies of the contingency plan should be delivered to office managers, and department and functional leaders.		

HYPOTHETICAL DISASTERS: COULD THESE HAPPEN TO YOU?

In each of the following situations, which have been adapted from actual events, you are the executive on the scene, in charge and you have total responsibility for results. There is no one "right," answer to any of these problems, and the situations address the three (3) types of incidents, caused by:

1. Natural or cataclysmic events (i.e., earthquake, fire, flood, storm).
2. Human behavior (i.e., robbery, bomb threat, arson, hostage event, transportation strike).
3. Technological breakdowns (i.e., power outage, computer crash). Analyze these situations and prioritize your response according to issues likely to affect:
    - People: employees and customers.
    - Places: facilities.
    - Things: assets and records.

NATURAL OR CATACLYSMIC EVENTS

Your company operates in only one facility, the two-story Main Office. The first floor is occupied by your customer services area, business operations and storage/inventory departments. The second floor is occupied by your administrative and data processing departments.

On a Saturday at 9:30 PM, an employee driving by the Main Office observes water gushing from under and between the front doors and into the street. You go to the facility to conduct a damage assessment, and discover that a water main on the second floor has broken, causing a water leak from the second floor into the first floor where your total inventory is located.

Your company has many facilities that are used by various business, operations and administrative functions. These facilities are strategically distributed throughout several communities according to your marketing plan.

On a Friday afternoon at 3:30 PM, the geographical area in which several offices are located suffers a catastrophic, community-wide hurricane. Fifteen employees and eight customers are trapped within an office and unable to leave for at least three days.

Police, fire and paramedic personnel are unavailable.

Telephone communications links for your area code have been destroyed; electrical service and water flow have been disrupted. Food and water supplies exist only within the building, and no deliveries are possible.

Your company services a remote resort community, and this community is accessed by one roadway. Many residents, employees, vacationers, and vendors must use this roadway to drive to the facility, which spans a river by means of a bridge.

Natural or Cataclysmic Events (CONT.)

There is a community-side" and a "resort-side" of the bridge. Businesses are located on the "community-side" and residences are located on the "resort-side."

On a Sunday at 2:00 AM the bridge collapses for unknown reasons. No one is injured, no buildings are damaged and all utilities remain operational.

Several of your company's departments occupy offices in a business park, as do many other businesses. The departments occupying the business park are your administration, finance, training and personnel departments.

Unknown to you, one of the other businesses located within this business park is illegally storing and shipping fireworks from its warehouse. On a Wednesday morning at 4:00 AM, a faulty water heater explodes within that business and burns the warehouse, causing the fireworks to explode and destroy the business park.

HUMAN-CAUSED EVENTS

Your company operates in only one facility, the two-story Main Office. The first floor is occupied by your customer services area, business operations and storage/inventory departments. The second floor is occupied by your administrative and data processing departments.

On a Monday afternoon at 2:00 PM, a truck driver driving an 18-wheeler on the street in front of the facility at 10 mph suffers a heart attack. He loses control of the truck, which carries two tanker-trailers of a flammable, toxic pesticide. The truck veers into the customer services area, shattering glass windows, and comes to rest against the customer service counter.

Your company has many facilities that are used by various business, operations and administrative functions. These facilities are strategically distributed throughout several communities according to your marketing plan. Your latest personnel survey reveals that 41% of your workforce is comprised of single parents of children under 12 years old.

Your local law enforcement agency receives numerous reports of sexual abuse at local privately-operated children's day care centers, and temporarily closes all such centers in your community to investigate. A significant number of your employees depend upon these centers to care for their children, and the employees must now stay home rather than report for work.

Your company operates in only one facility, the four-story Main Office. The first floor is occupied by your customer services area, operations and data processing departments. The second floor is occupied by your administration and finance departments. The third and fourth floors are leased to tenants.

A receiving clerk opens a mis-shipped carton to obtain re-shipping information on a Tuesday morning at 11:00 AM. Inside the box are several plastic baggies of a white powdery substance. The clerk tries to remove one of the baggies for inspection, and rips the corner of the baggie on the sharp edge of the box.

Human - Caused Events (CONT.)

The startled clerk drops the baggie, showering the work area with white powder which is immediately dispersed into the ventilation system. Several employees within the immediate area are immediately exposed to the powdery substance.

Your company has many facilities that are used by various business, operations and administrative functions. These facilities are strategically distributed throughout several communities according to your marketing plan. You are the office manager and arrive for work on a Friday morning at 8:00 AM, just minutes behind the first group of employees to enter the building. You look into the building from outside. You see that a gunman has taken two employees hostage, and is yelling to you, "Get inside, or I'll kill 'em."

Your company employs forty people and serves a community of 20,000 people. There are no other similar companies represented within this geographic area.

A food handler at a local fast-food restaurant is an intravenous drug user, and contracts infectious hepatitis. He then spreads this disease throughout the community serviced by your company. Your employees also live in this community, and several are now incapacitated for up to six months.

Your company has many facilities that are used by various business, operations and administrative functions. These facilities are strategically distributed throughout several communities according to your marketing plan. Several of these facilities are located within economically disadvantaged and crime-prone geographical areas.

Directly across the street from your company is another company in the midst of a labor strike. More than fifty people patrol the sidewalks carrying protest signs. In the past few days there have been periodic incidents of vandalism, violence and intimidating behavior. The situation is closely monitored by the news media. On Wednesday at 3:00 PM a car carrying several protestors drives by the company and throws a firebomb at the company's building. The immediate effect is an emergency callout of police and fire personnel who contain the area, the arrival of ambulances, and congestion caused by vehicles and pedestrian on-lookers.

Your company employs twenty-five people and serves a close-knit community of 20,000 people. All of your employees are within the facility, as are five customers and your spouse who is waiting to have lunch with you.

On a Monday morning at 11:45 AM, a man walks into the office and up to your desk. You recognize the man as the recently-divorced husband of one of your employees, Thelma, and you are aware that the husband has behaved violently and delivered death threats to Thelma in the past. The man is wearing a hunting vest with what appears to be several sticks of dynamite and wires protruding from the cartridge loops and pockets. The man says to you, "Bring Thelma out here. I want to talk to her. Do it now and you won't get hurt, and neither will she."

Your company has many facilities that are used by various business, operations and administrative functions. These facilities are strategically distributed throughout several communities according to your marketing plan.

Human - Caused Events (CONT.)

The company has been advertising the grand opening and dedication of its newest office, which has been operating for two weeks.

On a Thursday afternoon at 5:15 PM, your new office is holding the open house. All of your employees are present, as are several company executives and prominent members of the community. The PBX operator answers the phone as it rings, and hears: "Great party you're having. Because you didn't invite me, I planted a bomb in the office. The bomb's set to go off at 5:30, and I won't tell you where it is." Click....

Your up-scale, Highly-visible Company employs seventy-five people, and serves a metropolitan community of 220,000 people.

Your 9-year old child, Fred, walks to Lincoln Elementary School every school-day morning at about 7:30 AM. On a Wednesday morning at 7:45 AM, you receive this call: "We kidnapped your son, Fred, on his way to school. Today he's wearing **************). Get $25,000 in cash, no bills larger than $50s, put the money in a brown paper grocery sack and put the sack next to the Federal Express pickup box at northwest corner of Alameda and Terry. Don't call the police, don't let anyone know what you're doing, and don't be followed. Deliver the money, and we'll let Fred go in an hour: don't get the money, and you'll never see Fred again." Click....

TECHNOLOGICAL EVENTS

Your company employs forty people, maintains a board of directors comprised of six local residents, and serves a community of 20,000 people. There is no other similar company represented within this geographic area. The same president and executive vice president have managed your company for the past fifteen years.

Your company's president and executive vice president fly together to an important conference. On the return flight on a Tuesday morning the plane crashes, with no survivors. No succession of management agreement has ever been developed.

Your company maintains a mainframe computer to provide in-house data processing services to all departments within the company. Your company does not contract for any services with a service bureau, although it does store archived data at a remote location.

On a Thursday at 12:00 PM you are notified that every computer terminal within every location is frozen. Further investigation indicates that a virus has infected your mainframe computer, and that is has effectively destroyed both hardware components and software programs.

Your company maintains both a Finance and a Credit department and employs an executive to administer each department. Files containing each department's transactions, correspondence, marketing materials and client lists are stored on each department's personal computer. Each department's secretary is responsible for backing up his/her computer each Friday at 3:00 PM, and the secretary stores the backups in his/her desk.

The finance and credit department executives fail to report for work on Monday. Your investigation reveals that the executives' residences are vacant, and that they were engaged in a romantic affair. You also discover that the executives withdrew all available funds from their checking and savings accounts on Friday at 4:00 PM. You then discover that both departments' personal computers' hard drives have been erased, and that the backup files are missing.

[This page intentionally left blank]

DISASTER RECOVERY

Sample Procedures
410

[This page intentionally left blank]

DISASTER MANAGEMENT PROCEDURES
Table of Contents*

POLICIES & PROCEDURES **PAGE**

*The Disaster Management Procedures Table of Contents references section page numbers (lower right corner).

[This page intentionally left blank]

SAMPLE COMPANY **Page 1 of 7**
SOP # ______ Revision:______ **Prepared by:____________**
Effective Date:______________ **Approved by:____________**

Title: **DMP101 - DISASTER MANAGEMENT PLAN**

Policy: Develop and implement a written recovery plan based upon critical and accessory functions for each branch, facility, department and function within the company.

Purpose: The Disaster Management Plan is designed to assist the Company in effectively planning for, and acting during, emergencies that might otherwise impact the company's operations.

Scope: The purposes for developing and implementing this Disaster Management Plan are to determine the critical and accessory functions of the company. This involves a continuing assessment of the company's assets and liabilities, personnel, equipment and supplies and facilities.

Definition: For the purposes of this plan, a disaster is defined as any event or circumstance that may significantly impact the company's operations. This includes any sudden, unexpected occurrence, natural or human-caused, which poses a significant threat to the company's personnel, customers, facilities, assets, records, and its delivery of services.

Procedure:

1.0 DISASTER MANAGEMENT TEAM

1.1 The following personnel comprise the Disaster Management Team:

CO-CHAIRPERSON: ____

Responsible for: Board of Directors
Public and Media Relations
Shareholder and Stockholder Relations
Administration

Backup: ____

CO-CHAIRPERSON:

____

Responsible for:	Board of Directors Public and Media Relations Administration Personnel Insurance Accounting Data Processing, Administrative Functions

Backup: ______________________________

CO-COORDINATOR:

____

Responsible for:	Operations Security Telecommunications Data Processing, Operations Functions

Backup: ______________________________

CO-COORDINATOR:

____

Responsible for:	Operations Courier Services/Transportation Telecommunications Office Support

Backup: ______________________________

MEMBER: ______________________________

Responsible for:	Courier Services/Transportation Office Support Purchasing and Supplies

Backup: ______________________________

MEMBER: ______________________________

Responsible for:	Audit Compliance, Legal

Backup:

MEMBER: ____________________________________
Responsible for: Facilities
Equipment and Supplies
Transportation

Backup: ____________________________________

MEMBER: ____________________________________
Responsible for: Branch Operations #1

Backup: ____________________________________

MEMBER: ____________________________________
Responsible for: Branch Operations #2

Backup: ____________________________________

MEMBER: ____________________________________
Responsible for: Branch Operations #3

Backup: ____________________________________

MEMBER:
Responsible for: Disaster Records

Backup: ____________________________________

2.0 ANNUAL RISK ASSESSMENT AND EVALUATION

2.1 All potential risks to the Company shall be evaluated on at least an annual basis. This annual assessment shall be conducted by each Department Leader, and the written results and recommendations shall be forwarded to the Disaster Management Team Coordinators.

2.2 Upon receipt of these written results and recommendations, the Disaster Management Team Coordinators shall also issue a report of findings to the Disaster Management Team Chairpersons. This report shall include:

- An overview of the status of each department regarding compliance with policies and procedures.
- A projection regarding any new equipment, personnel, policies or procedures required for the next year, to meet the requirements of this plan.
- Any necessary change in current policies or procedures not addressed elsewhere, and an accompanying rationale or explanation.

2.3 In conducting this annual assessment, the following criteria shall be used by both Department Leaders and the Disaster Management Team Coordinators:

- Any potential risk shall be identified, whether already addressed or newly anticipated. If the identified risk has already been successfully addressed, no further assessment or recommendations shall be required. Any newly-identified risk shall be fully addressed.
- Identified risks may be categorized as natural or technological disasters, or caused by human actions. The listed occurrences have been identified as potentially creating an adverse impact upon the company's operations:
    - Fire
    - Power outage
    - Water leak
    - Gas leak
    - Bomb threat
    - Earthquake
    - Flood
    - Toxic (hazardous) waste dispersion
    - Civil disturbance
    - Bomb explosion
    - Sabotage
    - Tornado
    - Hurricane
    - Nuclear attack

2.4 Each of these potential risks shall be evaluated according to the following guidelines:

- What is your ability to predict this occurrence?
- What is the likely frequency of this occurrence?
- If applicable, what is the estimated elapsed time between your prediction and the onset of this occurrence?
- If applicable, what is the estimated elapsed time between your discovery of the occurrence and any injury or damage?

- What is the possibility of your controlling or diminishing the impact of this occurrence?
- What is the anticipated duration of this occurrence?
- What is the projected impact to the Company if this event occurs, regarding:
    - Personnel injury or working conditions.
    - Customer confidence and the company's image.
    - Security measures necessary to insure adequate protection of personnel, customers, and facilities, assets and records.
    - Legal issues surrounding the company's inability to process documents and other information.
    - Requirements to backup or substitute data or support services.
    - The effect upon related functions.
    - Recovery time.
    - Money lost.
    - Equipment lost.

2.5 Once potential risks have been identified and prioritized, the Disaster Management Team shall:

- Determine potential short and long range solutions.
- Decide upon the most appropriate primary and secondary solutions.
- Test and evaluate these solutions for effectiveness.
- Include these solutions in the Disaster Management Plan, and related testing and training programs.

2.6 All original records of the processes used to compile the information resulting in these policies and procedures shall be retained by the Disaster Management Team Coordinators, in a location separate from the Disaster Management Plan. These records shall include the names and positions of all appropriate persons involved in the continued planning and writing of this plan.

2.7 The following personnel are charged with the responsibility of the conception, development, initiation and maintenance of the Disaster Management Plan, as this plan pertains to their departments and/or responsibilities:

DEPARTMENT: Administration

OFFICER:

DEPARTMENT: Data Processing

OFFICER:

DEPARTMENT: Operations

OFFICER:

DEPARTMENT:

OFFICER:

DEPARTMENT:

OFFICER:

DEPARTMENT:

OFFICER:

DEPARTMENT:

OFFICER:

DEPARTMENT:

OFFICER:

DEPARTMENT:

OFFICER:

2.8 All insurance policies, and service and vendor agreements shall be reviewed at least annually. If policies and agreements are satisfactory, existing agreements shall be renewed. New vendors will be investigated to replace any unsatisfactory agreements. The listing of these agreements, emergency services agencies, and community resources shall be reviewed and updated at least annually.

2.9 The Disaster Management Team Coordinators shall also be responsible for researching and developing a listing of other product and service companies and individuals who may be utilized on an interim basis. This listing shall be developed within each county, or other appropriate geographic area, serviced by the company, to provide backup or restoration services until the Company has resumed normal operations. This listing shall be reviewed and updated at least annually.

[This page intentionally left blank]

TESTING OF SOLUTIONS DOCUMENTATION

On ________________________________the Disaster Management Plan solutions to the scenario indicated below were tested with the following results:

SCENARIO:

Test conducted by: ________________________________

Location/Department:________________________________

Date:________________________________Time:________________________________

Results:

____

____

____

____

____

____

____

EXHIBIT -1

[This page intentionally left blank]

SAMPLE COMPANY
SOP# ____ Revision:_____ **Prepared by:_________**
Effective Date:_________ **Approved by: ________**

Title: **DMP102 SERVICE AGREEMENTS, EMERGENCY SERVICES AGENCIES AND COMMUNITY RESOURCES**

Policy: The Disaster Management Team shall continually investigate, review and develop services to administer disaster preparation and recovery.

Purpose: To describe procedures required to prepare for disaster recovery.

Scope: This statement applies to disaster management team coordinators.

Procedure:

1.0 EMERGENCY SERVICES

The Disaster Management Team Coordinators shall continually investigate appropriate vendors and community resources to provide emergency services to the company. The Company shall rely, whenever possible, upon established vendors, internal devices, personnel and procedures to administer the recovery effort.

2.0 SERVICE AGREEMENTS

2.1 The Disaster Management Team Coordinators shall continue to seek appropriate service agreements with other companies, associations and service companies, to provide backup services until the Company has resumed normal operations. As those agreements are developed, they shall be added to the Disaster Management Plan, updated and verified at least annually.

2.2 The Disaster Management Team Coordinators shall also be responsible for developing and maintaining a comprehensive listing of all emergency services and related agencies within each county, or other appropriate geographic region, serviced by the company. Wherever possible, two (2) similar agencies, resources and vendors will be listed. This listing shall be updated and verified at least annually. (See following pages).

2.3 This listing shall focus upon those services and resources likely to assist with recovery efforts regarding:

- **Personnel:** Notification, basic needs, transportation, re-allocation, replacement, management, media relations, temporary lodging.
- **Customers:** Notification, marketing, public relations, cash-handling, credit approval.
- **Telecommunications**: Internal, external, public access, facsimile machine and modem, cellular.

- Emergency services: Law enforcement, fire, medical, non-emergency governmental agencies.
- Facilities: Power, utility agencies, repair and replacement contractors.
- Data Processing: Item processing, capture, proof, Fedwire, ACH, office PCs and peripherals.
- Equipment and supplies: Repair, replacement, special needs, delivery.

■ EMERGENCY SERVICES ■

Name: Police Department

City: ____

County: ____

Emergency Phone: 911

Business Phone: ____________________________________

Name: Sheriff's Department

County: ____

Emergency Phone: 911

Business Phone: ____________________________________

Name: Fire Department

City: ____

County: ____

Emergency Phone: 911

Business Phone: ____________________________________

Name: Paramedic/Rescue #1

City: ____

County: ____

Emergency Phone: 911

Business Phone: ____________________________________

Name: Air Ambulance

County: ____

Emergency Phone:

Business Phone: ____________________________________

Name: Private Ambulance #1

County: ____

Emergency Phone:

Business Phone: ____________________________________

Name: Private Ambulance #2

County: ____

Emergency Phone:

Business Phone:

HOSPITAL/URGENT CARE FACILITY:

Name: Hospital #1

Address:

City: ____

County: ____

Emergency Phone:

Business Phone: ________________________________

Name: Hospital #2

Address:

City: ____

County: ____

Emergency Phone:

Business Phone: ________________________________

Name: Hospital #3

Address:

City: ____

County: ____

Emergency Phone:

Business Phone: ________________________________

Name: Urgent Care Facility #1

Address:

City: ____

County: ____

Emergency Phone:

Business Phone: ________________________________

Name: Urgent Care Facility #2

Address:

City: ____

County: ____

Emergency Phone:

Business Phone: ________________________________

EMERGENCY STAGING FACILITIES AND SHELTERS

Name: American Red Cross

Address: ____

City: ___________________________________

County: _________________________________

Emergency Phone:

Business Phone: ____________________________

Staging location:

Name: Community Center #1

Address: ____

City: ___________________________________

County: _________________________________

Emergency Phone:

Business Phone: ____________________________

Staging location:

Name: Community Center #2

Address: ____

City: ___________________________________

County: _________________________________

Emergency Phone:

Business Phone: ____________________________

Staging location:

Name: National Guard Center

Address: ____

City: ___________________________________

County: _________________________________

Emergency Phone:

Business Phone: ____________________________

Staging location:

Name: Veterans Memorial Building

Address: ____

City:

County: ____

Emergency Phone:

Business Phone: ____________________________________

Staging location:

DISASTER INFORMATION:

Name: Medical Emergency Information Hotline

County: ____

Emergency Phone:

Name: Office of Emergency Services (City)

Address: ____

City: ____

County: ____

Emergency Phone:

Business Phone: ____________________________________

Staging location:

Name: Office of Emergency Services (County)

Address: ____

City: ____

County: ____

Emergency Phone:

Business Phone: ____________________________________

Staging location:

Name: Office of Emergency Services (State)

Address: ____

City: ____

County: ____

Emergency Phone:

Staging location:

COMMUNICATIONS:

Name: Company for remote contact #1

Address: ______________________________

Phone 1:

Phone 2:

FAX:

Name: Company for remote contact #2

Address: ______________________________

Phone 1:

Phone 2:

FAX:

Name: Mobile telephone #1

Address: ______________________________

Phone 1:

Phone 2:

Cellular access numbers:

Name: Mobile telephone #2

Address: ______________________________

Phone 1:

Phone 2:

Cellular access numbers:

Name: Newspaper #1

City: ______________________________

County: ______________________________

Emergency Phone:

Business Phone: ______________________________

FAX:

Name: Newspaper #2

City: ______________________________

County: ______________________________

Emergency Phone:

Business Phone:

FAX:

Name: Public telephone #1 (3 for each facility)

Address: ___________________________________

Phone:

Name: Public telephone #2 (3 for each facility)

Address: ___________________________________

Phone:

Name: Public telephone #3 (3 for each facility)

Address: ___________________________________

Phone:

Name: Radio Station #1

City: ___________________________________

County: ___________________________________

Emergency Phone:

Business Phone: ___________________________________

FAX:

Name: Radio Station #2

City: ___________________________________

County: ___________________________________

Emergency Phone:

Business Phone: ___________________________________

FAX:

Name: Television Station #1

City: ___________________________________

County: ___________________________________

Emergency Phone:

Business Phone: ___________________________________

FAX:

Name: Television Station #2

City: ___________________________________

County: ___________________________________

Emergency Phone:

Business Phone:

FAX:

SECURITY:

Name: Alarm Company #1

City: ____

County: ______________________________________

Emergency Phone:

Business Phone: ________________________________

Name: Alarm Company #2

City: ____

County: ______________________________________

Emergency Phone:

Business Phone:

Name: Guard Company #1

City: ____

County: ______________________________________

Emergency Phone:

Business Phone: ________________________________

Name: Guard Company #2

City: ____

County: ______________________________________

Emergency Phone:

Business Phone: ________________________________

TRANSPORTATION:

Name: Airport/Municipal (City)

Address: ____

City: ____

County: ______________________________________

Emergency Phone:

Business Phone: ________________________________

Name: Airport/Regional (County)

Address: ____

City: ___________________________________

County: ___________________________________

Emergency Phone:

Business Phone: ____________________________

Name: Bus (City)

City: ___________________________________

County: ___________________________________

Emergency Phone:

Business Phone: ____________________________

Name: Bus (County)

City: ___________________________________

County: ___________________________________

Emergency Phone:

Business Phone: ____________________________

Name: Cab/Taxi #1

City: ___________________________________

County: ___________________________________

Emergency Phone:

Business Phone: ____________________________

Name: Cab/Taxi #2

City: ___________________________________

County: ___________________________________

Emergency Phone:

Business Phone: ____________________________

Name: Car Rental Agency #1

City: ___________________________________

County: ___________________________________

Emergency Phone:

Business Phone: ____________________________

Name: Car Rental Agency #1

City:

County:

Emergency Phone:

Business Phone: ____

Name: Cash Transport/Supplemental #1

City: ____

County: ____

Emergency Phone:

Business Phone: ____

Name: Cash Transport/Supplemental #2

City: ____

County: ____

Emergency Phone:

Business Phone: ____

Name: Moving Company #1

City: ____

County: ____

Emergency Phone:

Business Phone:

Name: Moving Company #2

City: ____

County: ____

Emergency Phone:

Business Phone: ____

CITY/COUNTY OFFICES:

Name: Administrative Offices (City)

City:

County:

Emergency Phone:

Business Phone: ____

Name: Administrative Offices (County)

City:

County:

Emergency Phone:

Business Phone: ________________________________

Name: Air Quality Control Offices (County)

City: ________________________________

County: ________________________________

Business Phone: ________________________________

Business Phone:

Name: Animal Control (City)

City: ________________________________

County: ________________________________

Emergency Phone:

Business Phone: ________________________________

Name: Animal Control (County)

City: ________________________________

County: ________________________________

Emergency Phone:

Business Phone: ________________________________

Name: Building Inspector (City)

City: ________________________________

County: ________________________________

Emergency Phone:

Business Phone: ________________________________

Name: Building Inspector (County)

City: ________________________________

County: ________________________________

Emergency Phone:

Business Phone: ________________________________

Name: Hazardous Materials Team (County)

City: ________________________________

County: ________________________________

Emergency Phone:

Business Phone: ___________________________________

Name: Health (City)

County: ___________________________________

Emergency Phone:

Business Phone: ___________________________________

Name: Health (County)

City: ___________________________________

County: ___________________________________

Emergency Phone:

Business Phone: ___________________________________

Name: Health and Safety Inspector (County)

City: ___________________________________

County: ___________________________________

Emergency Phone:

Business Phone: ___________________________________

Name: Mental Health (City)

City: ___________________________________

County: ___________________________________

Emergency Phone:

Business Phone: ___________________________________

Name: Mental Health (County)

City: ___________________________________

County: ___________________________________

Emergency Phone:

Business Phone: ___________________________________

Name: Parks and Recreation (City)

County: ___________________________________

Emergency Phone:

Business Phone: ___________________________________

Name: Parks and Recreation (County)

County: ___________________________________

Emergency Phone:

Business Phone: ____

Name: Public Information Office (City)

City: ____

County: ____

Emergency Phone:

Business Phone: ____

Name: Public Information Office (County)

City: ____

County: ____

Emergency Phone:

Business Phone: ____

Name: Public Works (City)

City: ____

County: ____

Emergency Phone:

Business Phone:

Name: Public Works (County)

City: ____

County: ____

Emergency Phone:

Business Phone:

Name: Schools Information (City)

City: ____

County: ____

Emergency Phone:

Business Phone:

Name: Schools Information (County)

City: ____

County: ____

Emergency Phone:

Business Phone:

Name: Transportation/Roads (City)

City: ____

County: ____

Emergency Phone:

Business Phone:

Name: Transportation/Roads (County)

City: ____

County: ____

Emergency Phone:

Business Phone:

FEDERAL OFFICES:

Name: Department of Corporations

Address: ____

Emergency Phone:

Business Phone:

Name: Federal Bureau of Investigation

Address: ____

Emergency Phone:

Business Phone:

Name: Federal Reserve Bank

Address: ____

Emergency Phone:

Business Phone:

Name: Federal Reserve Board of Governors

Address: ____

Emergency Phone:

Business Phone:

Name: National Credit Union Administration

Address: ____

Emergency Phone:

Business Phone:

Name: Office of the Comptroller of the Currency

Address: ____

Emergency Phone:

Business Phone:

Name: Office of Thrift Supervision

Address: ____

Emergency Phone:

Business Phone:

Name: State Banking Department

Address: ____

Emergency Phone:

Business Phone:

Name: The Federal Deposit Insurance Corporation

Address: ____

Emergency Phone:

Business Phone:

UTILITIES:

Name: Electric

City: ______________________________________

County: ____________________________________

Emergency Phone:

Business Phone:

Name: Gas

City:

County: ____________________________________

Emergency Phone:

Business Phone:

Name: Sanitation/Sewer

City: ______________________________________

County: ____________________________________

Emergency Phone:

Business Phone:

Name: Telephone

City: ______________________________

County: ______________________________

Emergency Phone:

Business Phone:

Name: Water

City: ______________________________

County: ______________________________

Emergency Phone:

Business Phone:

CONTRACTORS/VENDORS:

Name: Building/Reconstruction Contractor

City: ______________________________

County: ______________________________

Emergency Phone:

Business Phone:

Name: Electrical Contractor

City:

County: ______________________________

Emergency Phone:

Business Phone:

Name: Glass Contractor

City: ______________________________

County: ______________________________

Emergency Phone:

Business Phone:

Name: Grounds keeping Contractor

City: ______________________________

County: ______________________________

Emergency Phone:

Business Phone:

Name: Hazardous Materials Response Contractor

City: ______________________________

County: ______________________________

Emergency Phone:

Business Phone:

Name: Health and Safety Contractor

City: ______________________________

County: ______________________________

Emergency Phone:

Business Phone:

Name: Janitorial Contractor

City: ______________________________

County: ______________________________

Emergency Phone:

Business Phone:

Name: Locksmith and Vault Contractor

City: ______________________________

County: ______________________________

Emergency Phone:

Business Phone:

Name: Office Supplies and Forms Contractor

City: ______________________________

County: ______________________________

Emergency Phone:

Business Phone:

Name: Plumbing Contractor

City: ______________________________

County: ______________________________

Emergency Phone:

Business Phone:

Name: Roofing Contractor

City:

County: ____

Emergency Phone:

Business Phone:

Name: Utility Contractor

City: ____

County: ____

Emergency Phone:

Business Phone:

EQUIPMENT:

Name: Applications

City: ____

County: ____

Emergency Phone:

Business Phone:

Name: Mainframe computers

City: ____

County: ____

Emergency Phone:

Business Phone:

Name: Office equipment other than PCs

City: ____

County: ____

Emergency Phone:

Business Phone:

Name: Personal computers

City: ____

County: ____

Emergency Phone:

Business Phone:

Name: Software

City: ____

County: ____

Emergency Phone:

Business Phone:

SAMPLE COMPANY **Page 1 of 2**

SOP # ______ Revision:______ **Prepared by:____________**
Effective Date:______________ **Approved by:____________**

Title: **DMP103 CENTERS OF OPERATION**

Policy: The Company will provide for a primary and secondary centers of operation equipped with appropriate supplies and equipment.

Purpose: To prepare for safety and recovery as a result of a disaster.

Scope: This statement will apply to alternative centers of operations and recommended supplies.

Procedure:

1.0 SUPPLIES AND EQUIPMENT

The following supplies and equipment shall be stored at the primary and secondary Centers of Operations, and at each branch, department and facility as appropriate:

- Current full copy of the company's Disaster Management Plan and operations procedures for each branch, department, facility and function.
- Building plans for each facility, including primary control points for all utilities, if applicable.
- Complete and current list of all personnel, including names, addresses, and phone numbers. Current position descriptions indicating routine duties and responsibilities shall be included.
- Complete and current list of all equipment and serial numbers; replacement information; names and phone numbers of designated vendors; and service agreement information.
- Complete and current list of all emergency services, community resources and related agencies within each county, or other geographic region, serviced by the company.
- Food, water and medical supplies as appropriate.
- Battery-powered radios capable of receiving emergency instructions; fire extinguishers; flashlights; and basic tools as appropriate.

CENTERS OF OPERATION **Page 2 of 2**

- Master keys, alarm access codes and other security information for each branch, facility and the other Centers of Operations.
- Stationery items, numbered forms and negotiable documents as appropriate.
- Assorted writing materials, instruments and business tools as appropriate.
- Non-electrical mechanical devices as appropriate.
- Chronological log forms to document all actions taken during the emergency.
- Any other items considered appropriate during the initial and annual review process.

2.0 STORAGE SITE

In addition to the primary and secondary Centers of Operations, a storage site may be maintained at a location distant from those Centers. This storage site may contain items acquired in addition or in place of the items stored at the Centers of Operations. These items are to be determined by the Disaster Management Team Coordinators. This storage site is:

Address: ______________________________________

City: ____

State: ZIP:

Phone: FAX:

Space #:

Keys are maintained by:

SAMPLE COMPANY **Page 1 of 3**

SOP # ______ Revision:______ **Prepared by:_____________**
Effective Date:______________ **Approved by:_____________**

Title: **DMP104 EMERGENCY NOTIFICATION PROCEDURES**

Policy: The information within this section shall be maintained and updated by the Disaster Management Team Chairpersons, Coordinators and Department Leaders.

Purpose: To provide required procedures for emergency notification.

Scope: This statement applies to all disaster management team members.

Procedure:

1.0 EMPLOYEE LISTS

1.1 The Disaster Management Team Chairpersons, Coordinators and Department Leaders shall maintain appropriate lists of employees' names, titles, position descriptions, home addresses and home telephone numbers. This information shall be maintained with the Disaster Management Plan for contact, if necessary.

1.2 This personnel information shall be updated at least annually, and shall remain confidential. (Reference Ex1, EMPLOYEE PERSONAL PROFILE)

2.0 NOTIFICATION AND RESPONSE

2.1 Upon notification of an emergency, the Disaster Management Team Chairpersons shall authorize the notification and immediate response of appropriate personnel. This notification shall be made by phone or in person, and according to one of the three levels of threat as follows:

1. **ALERT 3: STAND-BY:** Appropriate personnel are to be notified of the emergency and placed on stand-by status, and shall remain available for emergency assignment, either by phone or by other means as necessary. If the emergency occurs during business hours:
    - All personnel are placed on alert.
    - All appropriate records, equipment and documents are prepared for transfer to secure areas or facilities.
    - All operations may resume routine functions.

2. **ALERT 2: MINIMUM RESPONSE**: Only selected personnel will be called to respond immediately, and all other personnel will remain on stand-by status. If the emergency occurs during business hours:

    - All personnel are to remain at current locations, until instructed to leave.
    - All appropriate records, equipment and documents are transferred to secure areas or facilities; and business operations involving customers are discontinued, and internal tasks, functions and assignments continue until further instructions are received.

3. **ALERT 1: MAXIMUM RESPONSE:** All designated personnel shall respond

    The Company or any of its facilities may be closed immediately, and all personnel are to remain at their current locations.

    - All appropriate records, equipment and documents are locked within a secure facility.
    - All business operations are discontinued, and personnel and customers remain at their locations until escorted from those locations by company officials or emergency services personnel.

2.2 In considering the appropriate response, the Disaster Management Team shall consider which stage(s) of emergency is present:

- Detection
- Reaction
- Assessment
- Notification
- Mobilization
- Recovery
- Restoration
- Resumption
- Reconstruction

3.0 ACTIVITY LOG

In the event of any activation of this plan, the Disaster Management Team Coordinators shall cause an activity log to be kept at all times during the activation. This log shall be maintained in a chronological sequence of events, and clearly identify the following information:

- Time activity begun, actions taken, activity terminated.
- Locations involved.
- Person maintaining log entries.
- Person generating information.
- Person assigned a task, if applicable.
- Notification of other personnel, if applicable:
    - Name of person notified
    - Time of notification
    - Action requested
    - Disposition
- Unusual events requiring follow-up when normal operations resume.
- Termination date/time of emergency activation, including name and title of person making termination.
- Reference **EXHIBIT 2, CHRONOLOGICAL LOG OF EVENTS**)

[This page intentionally left blank]

EMPLOYEE PERSONAL PROFILE

The information requested in this questionnaire is voluntary and confidential, and is not to be used for any purpose other than during an actual emergency. The contents of this questionnaire shall be kept in a sealed envelope in a secure area, and shall not be opened unless in the case of an actual emergency. The contents of this questionnaire and a photograph shall be updated annually during your performance evaluation.

PERSONAL IDENTIFYING INFORMATION:

Employee: ______________________________

Nicknames or other names used: ______________________________

Employment classification: ______________________________

Employment location: ______________________________

Permanent residence: ______________________________

______________________________ Phone: ______________

Secondary residence: ______________________________

______________________________ Phone: ______________

Other employment, if applicable: ______________________________

Date of birth: / / Place of birth: ______________________________

Name of hospital: ______________ Mother's name: ______________

Race: ______ Sex: ______ Complexion: ______________

Height: _____ Weight: _____ Hair color: ____________ Eye color: ____________

Scars/marks/tattoos: ______________________________

Hobbies: ______________________________

Are your fingerprints on file with the company? ______________

Is a current photograph on file with the company? ______________

EXHIBIT 1

FAMILY AND EMERGENCY NOTIFICATION INFORMATION

Marital status: ________________________ Anniversary date: ____________________
Name of spouse/roommate: _____________ Nickname: _________________________
Number of children: __________________ Children's names are:
____ Birthdate: _________
____ Birthdate: _________
____ Birthdate: _________
____ Birthdate: _________

PERSONS TO CONTACT IN CASE OF EMERGENCY

Name: ____ Phone: _______________
Address: ____ Relationship: ___________
Name: ____ Phone: _______________
Address: ____ Relationship: ___________
Name: ____ Phone: _______________
Address: ____ Relationship: ___________

IMMEDIATE CLOSE RELATIVES

Name: ____ Phone: _______________

Address: ____ Relationship: ___________

Name: ____ Phone: _______________

Address: ____ Relationship: ___________

Name: ____ Phone: _______________

Address: ____ Relationship: ___________

EXHIBIT 1

OTHER PERSONS LIVING OR WORKING IN HOUSEHOLD

Name: ______________________________________ Phone: ____________
Address: ____________________________________ Relationship: _________
Name: ______________________________________ Phone: ____________
Address: ____________________________________ Relationship: _________

MOTOR VEHICLES USED BY FAMILY

Year: _______ Make: _____________ Model: ____________ Color: ___________
License: ______________________ Driven by: ___________________________
Year: _______ Make: _____________ Model: ____________ Color: ___________
License: ______________________ Driven by: ___________________________

MEDICAL INFORMATION

Blood type: ________ Allergic to: ____
Medical condition requiring treatment or medication:

____

____

Medication required: _____________________ Dosage: ____________________
Personal physician: ______________________ Phone: _____________________
Address: ____

AUTHORIZATION FOR THIS PHYSICIAN TO RELEASE CONFIDENTIAL INFORMATION IN THE EVENT OF AN EMERGENCY SITUATION REQUIRING TREATMENT:

Signature: _____________________________ Date: _______________________

EXHIBIT 1

[This page intentionally left blank]

CHRONOLOGICAL LOG OF EVENTS				
DATE:________________TIME STARTED: ______________TIME ENDED:________				
LOCATION:____				
OFFICER IN CHARGE:______________________ OPERATOR:__________________				
Time	Person	Activity/request	Disposition	Time

EXHIBIT 2

[This page intentionally left blank]

SAMPLE COMPANY
SOP # ______ Revision:______ **Prepared by:____________**
Effective Date:______________ **Approved by:____________**

Title: **DMP105 OFFICE AND DEPARTMENT RECOVERY**

Policy: In the unfortunate occurrence of a disaster, the following procedures shall be strictly followed.

Purpose: To describe critical functions required to maximize safety and minimize recovery time.

Scope: This statement applies to all management team members and employees.

Procedure:

1.0 TELEPHONE USE

1.1 Maintaining contact with members of the company's Disaster Management Team during a disaster is critical to a successful recovery effort. Usual business phone numbers are listed below, and these numbers should be used for all primary contact with Team members.

Disaster Management Chairpersons:

Name:___________________________ Phone: _______________

Emergency Phone: ___________________

Cellular Phone: _____________________

Name:___________________________ Phone: _______________

Emergency Phone: ___________________

Cellular Phone: _____________________

Disaster Management Coordinators:

Name:___________________________ Phone: _______________

Emergency Phone: ___________________

Cellular Phone: _____________________

Name:___________________________ Phone: _______________

Emergency Phone: ___________________

Cellular Phone: _____________________

Facilities/Equipment/Supplies/

Transportation/Telecommunications:

Name: Phone: _________________

Emergency Phone:

Cellular Phone:

Other Leaders:

Name:___________________________ Phone: _______________

Emergency Phone: __________________

Cellular Phone: ____________________

Name:___________________________ Phone: _______________

Emergency Phone: __________________

Cellular Phone: ____________________

Name:___________________________ Phone: _______________

Emergency Phone: __________________

Cellular Phone: ____________________

Name:___________________________ Phone: _______________

Emergency Phone: __________________

Cellular Phone: ____________________

Name:___________________________ Phone: _______________

Emergency Phone: __________________

Cellular Phone: ____________________

Name:___________________________ Phone: _______________

Emergency Phone: __________________

Cellular Phone: ____________________

1.2 In the event that normal phone lines are not functional, alternate communications may be available by using public phones (Bell system only). The public phones most readily accessible by company personnel are:

Public Phone # 1: -_____________________________

Location: _____________________________________

Public Phone # 2: _______________________________

Location: _____________________________________

Public Phone # 3: _______________________________

Location: _____________________________________

1.3 The Company has established a hotline phone number for emergency use by all employees. Use of this special number is restricted to disaster recovery efforts and emergency notifications only, and is not to be used for any other purpose.

Disaster Recovery Hotline number for employees only:

Security Alarm Company:

2.0 IMMEDIATE ACTIONS

The Leader or designate is to immediately take the following actions if a disaster occurs:

- Assess any injury or damage to employees, customers, and facilities.
- Temporarily close and secure the facility, if necessary.
- Contact appropriate emergency services, if necessary.
- Begin documenting the effects of the disaster and actions taken; secure all assets and records.
- First attempt to contact the company's Disaster Management Team Coordinators or Chairpersons with a damage assessment and actions taken report, and act upon instructions received.
- If all documented attempts to communicate with Coordinators and Chairpersons have failed, the Leader or designate is authorized to initiate reasonable and prudent responses necessary to minimize potential:
    - Injuries to personnel and customers
    - Damage to facilities
    - Loss of assets and records

3.0 COMPANY FUNCTIONS

3.1 The critical functions of the Company are to be serviced before the performance of any other task. The critical functions are:

- Administrative operations.
- Business operations.
- Board of Directors.
- Media relations.
- Security.
- Stock/shareholder relations.

3.2 The accessory functions of the Company are to be addressed only after all critical functions have been performed. The accessory functions are:

- Personnel.
- Accounting/Auditing.
- Accounts payable.
- Maintain custody of important company records.
- Insurance policies and Board-related matters.

4.0 DUTIES AND RESPONSIBILITIES

4.1 Description of the Leader's duties and responsibilities during a disaster:

- Ultimately responsible for overall company operations, including all personnel, customers, facilities and assets; investments, insurance, marketing, security, and company policies and procedures.
- Disaster Management Team Co-Chairperson.
- Ensure adequate supervision for all personnel and functions while absent from the Company or unavailable for contact; operational quality control.
- Respond to and comply with all regulations, policies and procedures regarding Department operations; conduct inquiries, interviews and investigations, and prepare written reports of findings as required.
- Public, media and shareholder relations; resolving customer issues where appropriate; approve unusual or unique transaction when no other person has that immediate authority; provide information to media representatives, and approve all requests for interviews from the press, radio and television.
- Provide appropriate members of the Disaster Management Team with accurate and timely information updates regarding the company's recovery efforts.
- Other duties and responsibilities as required.

4.2 Description of the Assistant Leader's duties and responsibilities:

- Perform all duties and responsibilities of the Department Leader in his/her absence or because of his/her unavailability.
- Administrative Security Officer, Administrative Compliance Officer and Department Disaster Management Team Co-Chairperson.
- Manage day-to-day administrative functions, oversees the company's maintenance of adequate insurance; and directly supervise appropriate personnel.
- Other duties and responsibilities as required.

5.0 EVACUATION

If a disaster occurs during working hours, the staff will evacuate the facility and assemble at:

PRIMARY:

Location: ______________________________

SECONDARY:

Location: ______________________________

6.0 ALTERNATE SITES

If the Company is not able to be used, operations will immediately shift to these alternate sites:

PRIMARY:

Location: ______________________________________

Address: ______________________________________

City: ____

State: ZIP:

Phone: ________________________

FAX: __________________________

SECONDARY:

Location: ______________________________________

Address: ______________________________________

City: ____

State: ZIP:

Phone: ________________________

FAX: __________________________

7.0 RECOVERY

7.1 If the Company is still operable, this checklist describes the functions or sections upon which you will concentrate recovery efforts, and in what order. Before opening the Company:

- Assess safety considerations for employees and customers.
- Coordinate with emergency services agencies, if necessary.
- Conduct a damage assessment of the building; determine levels of operation or restoration time for electricity, telephones, water, computer system.
- Ensure all areas of responsibility are staffed.
- Ensure equipment and supplies are available.
- Arrange for the safe relocation of all records and equipment.

7.2 If it is safe to open the Company, re-establish:

- Employee, customer, facility, assets and records security.
- Contact with the Board of Directors.
- Contact with media representatives.

- Corporate files and financial records.
- Personnel and fixed asset records.
- Accounts payable records.
- Insurance policies and documentation.

7.3 The Company requires these logistical factors to be available to perform critical functions:

- Square feet:
- Maximum number of employees:
- Maximum number of customers:
- Special relocation needs in the event the facility is unable to support Department operations are:

RECOVERY WORKSHEET FOR: ______________________________

Following the onset of an emergency or disaster requiring the activation of the company's Disaster Management Plan, the following procedures must be implemented to recover/stabilize operations:

1. Follow emergency instructions for specific position descriptions, addressing employee and customer safety first.
2. Assess any injury to employees and customers, and damage to facilities.
3. Temporarily close and secure the facility, if necessary.
4. Contact appropriate emergency services, if necessary.
5. Begin documenting the effects of the disaster and actions taken; secure all assets and records.
6. First attempt to contact the Disaster Management Team Coordinators with a damage assessment and actions taken report, and act upon instructions received. If the Coordinators are unavailable, contact Chairpersons for instructions.
7. If all documented attempts to communicate with Coordinators and Chairpersons have failed, the Department/Branch Leader or designate is authorized to initiate reasonable and prudent responses necessary to minimize potential:
    1. Injuries to personnel and customers.
    2. Damage to facilities.
    3. Loss of assets and records.
8.
9.
10.

EXHIBIT 1

[This page intentionally left blank]

RESTORATION WORKSHEET FOR: ______________________________

Following the recovery/stabilization of operations, the following procedures must be implemented to restore minimum levels of service:

1. Based upon the actual results of the disaster, review and temporarily revise the disaster response plan appropriately.
2. Assign the performance of critical tasks to designated employees, and initiate and maintain reporting structure and requirements.
3. Conduct a damage assessment of all equipment and supplies. Arrange for the transportation of replacement equipment and supplies, if appropriate.
4. Verify that all necessary equipment and supplies are available to perform critical tasks.
5. Retrieve data files, vital records, and other critical work.
6. Provide for physical site security, and institute other security policies and procedures.
7. Clear the work areas and provide temporary repairs as appropriate.
8. Continue documenting the effects of the disaster, and all actions taken to restore services. Photograph the impact areas if possible.
9.
10.
11.
12.

EXHIBIT 2

[This page intentionally left blank]

RESUMPTION WORKSHEET FOR: ______________________________

Following the restoration of basic operations, the following procedures must be implemented to resume normal levels of service:

1. Verify that basic operations have been fully restored, and notify the Disaster Management Team Coordinators of the completed restoration.
2. Determine personnel, equipment and supply needs to complete the restoration of normal services.
3. Replace or restore equipment as appropriate.
4. Continue documenting the effects of the disaster, and all actions taken to restore services.
5.
6.
7.
8.
9.
10.
11.
12.

EXHIBIT 3

[This page intentionally left blank]

RECONSTRUCTION WORKSHEET FOR: ______________________________

Following the resumption of normal operations, the following procedures must be implemented to reconstruct this department to the level of operation existing prior to the emergency:

1.
2.
3.
4.
5.
6.
7.
8.
9.
10.
11.
12.

EXHIBIT 4

[This page intentionally left blank]

Sample Memos, Agreements and Forms 420

[This page intentionally left blank]

SAMPLE MEMOS, AGREEMENTS AND FORMS

Table of Contents*

* The Sample Memos, Agreements and Forms Table of Contents references section page numbers (lower right corner).

3.0 FORMS

4.0 TRAINING HANDOUT

To: All Disaster Management Team Members

From:

Date:

Re: Disaster Management and Business Recovery Plan

The company has made a commitment to research, develop and implement a comprehensive Disaster Management Plan for each office, department, and facility. You have been selected to assist in this effort as a member of the Disaster Management Team.

The company has granted me the authority to oversee the project, to delegate and monitor assignments to all Team members, and the final responsibility for results. The Disaster Management Plan is to be completed within a time frame to be discussed at our first meeting.

Attached is the initial draft of your Disaster Recovery Questionnaire (Section 16). The information requested is critical to developing the company's overall Disaster Management Plan.

Each Office and Department Leader, and other selected personnel, will be directed to complete additional questionnaires and assessments by the Disaster Management Team Chairpersons or me.

When you have received this information, please review its contents carefully and be prepared to discuss your questions and concerns at our first meeting. Do not begin researching the information until we've had the opportunity to discuss the best methods, and any special considerations.

I will advise each Team member of the date, time and location of our first meeting. The purpose of this initial meeting is as follows:

- Assign the personnel to be responsible for the research, development and implementation of the Disaster Management Plan for the company.
- Discuss the purpose, procedures and related topics necessary for the successful completion of the Disaster Management Plan. This may require a brief review of the Disaster Management Plan's sections and scope.
- Assign tasks and time-lines for the completion of all phases of the Disaster Management Plan, and address any questions.
- This encompasses all offices, departments, facilities, functions and personnel.
- Develop the reporting and coordination routines for the gathering and processing of information.

EXHIBIT 1 - 1

After our initial meeting:

Select an assistant if appropriate, and brief him/her regarding the purpose and format of this assessment and questionnaire.

Assign the research and development of appropriate portions of this information to an employee, if possible.

Set and maintain a date for the final effort to be completed, and then personally review the compiled information for completeness and legibility.

Forward the completed questionnaire to me as assigned, so that your department's efforts are combined with others', and a complete evaluation may be sent to the Disaster Management Team Chairpersons.

Thanks for your efforts, and please contact me directly with your questions.

EXHIBIT 1 - 2

To:

From: Disaster Management Team Chairpersons

Date:

Re: Annual Risk Assessment Exercise

In preparation for the annual Disaster Management Plan review, ensure that the listed tasks have been accomplished within the next thirty (30) days:

1. Disaster Management Team Coordinators will circulate the current copy of the recovery script to all offices and departments for review. Within ten (10) days, Department and Office Leaders will note any changes, and return for edit, reprint and redistribution, if necessary.
2. The company's managers responsible for insurance and facilities will review all appropriate contracts and insurance policies, and advise the Disaster Management Executive Committee of any necessary changes.
3. Office and Department Leaders will review all appropriate training and documentation devices to insure compliance with local, state and federal regulations, and advise the Disaster Management Executive Committee of any necessary changes.
4. The company's auditor will review, by questionnaire or in person, all offices, departments and facilities for compliance with all regulatory requirements, and advise the Disaster Management Executive Committee of any negative findings.
5. Disaster Management Executive Committee will review any proposed changes or other developments affecting the plan, and direct necessary changes.
6. A synopsis of changes made or other developments will be presented to the Board of Directors. The Board will then re-certify the plan, with this action reflected in the Board's minutes.
7. Training personnel will revise training materials as appropriate, and conduct update training for offices, departments and facilities.

EXHIBIT 2

[This page intentionally left blank]

To: Office and Department Managers

From:

Date:

Re: Disaster Management Plan: Annual Testing and Training Program

Attached are three (3) items for placement in your copies of the Disaster Management Plan for your office or department:

Recovery script and emergency staging area diagrams.

Emergency services and disaster recovery information in database printout form. Remember that much of this information is confidential, and must be destroyed as any other confidential document.

Training handout, leader's guide and practical problem-solving exercise.

Deliver your annual testing and training meeting within the next ten (10) days, and:

Complete the Testing of Solutions Documentation;

Sign off the certification memo;

Make a copy of the certification memo and testing report for your files, and return the originals to me immediately following the training meeting; and

Keep the copy of the memo and report, the training handout and leader's guide with your office/department plan binder.

Thanks for your efforts, and please call me directly with any questions.

EXHIBIT 3

[This page intentionally left blank]

To: Board of Directors

From:

Date:

Re: Disaster Management and Business Recovery Plan Orientation

The purpose of this memo is to provide you, as a member of the Board of Directors, with necessary information regarding your role in implementing the company's Disaster Management and Business Recovery Plan.

The attached material contains:

- An overview of the plan, its goals and objectives, and the rationale for its implementation;
- A description of the duties and responsibilities assigned to various groups within the company;
- Listings of appropriate locations, contact personnel and telephone numbers; and
- Guidelines for the company's personnel and Board members to follow if the plan is activated.

I have provided you with two (2) copies of this material. Please store one copy at your residence, and the other in a convenient location, such as the trunk of your vehicle.

Please contact me directly with any questions, and thank you for your efforts in helping the company create a pro-active disaster recovery plan.

EXHIBIT 4

[This page intentionally left blank]

To: Disaster Management Team Chairpersons

From: Disaster Management Plan Coordinators

Date:

Re: Synopsis, Disaster Management Plan

The completed Disaster Management Plan for the company is submitted to you for approval by the Board of Directors. This memo is to serve as a synopsis of the plan.

Sections 01 through 03 address the procedure for the formulation of the company's policy and procedure mechanism relating only to the plan: an overview to the concept of disaster planning; and the company's statement of purpose regarding emergency activation of any of the measure included in the plan.

Sections 04 through 6.7 identify all personnel responsible for the initial and continuing research, development and implementation of the plan; establishes the formal notification and reporting structure for all members of the Disaster Management Team; describes the duties and responsibilities of all personnel; and assigns appropriate levels of authority to those personnel.

Section 7 described the alert scenario for differing levels of emergency; and the stages of emergency used to calculate the most effective response.

Section 8 identifies the Emergency Response Procedures that all employees must be able to identify, prioritize and act upon to diminish the impact of common events.

Section 9 addresses agreements for services and agencies responsible for coordinating disaster management and business recovery efforts in the community.

Section 10.0 identifies the site of primary and secondary Centers of Operations, to be used in the event this plan is activated; an itemized list of supplies and equipment to be distributed to these Centers of Operations; an itemized list of supplies and equipment to be stored at each office; and an itemized list of supplies and equipment to be stored at the storage site, if appropriate.

Section 11.0 describes the testing and training requirements to be developed upon the Board's approval of the plan. This testing and training program has been developed to address the company's entire staff regarding policies, procedures, equipment and other topics to be developed.

EXHIBIT 5 - 1

Section 12.0 describes the annual evaluation process to be conducted by each department. The annual report, to be issued by the Disaster Management Team Coordinators and approved by the Board, must address the status of each department regarding compliance with established policies and procedures; a projection regarding any new equipment, personnel, policies or procedures required for the coming year; and suggested changes to existing policies or procedures with accompanying rationale.

Sections 13.0 through 16.0 contains each department's and office's recovery plan for resumption of normal services after a disaster. This section includes identifying information on all personnel, equipment, facilities and supplies; disaster evacuation, assembly and notification procedures; and communications information.

Section 17 contains the training program, to be supplemented with vendor and related information as appropriate.

Section **18** is reserved for information added to the plan and documented changes at intervals.

Section 19 contains the forms to be used by the company during an emergency activation of the plan.

Section 20 is the actual resolution authorized by the Board, and insures the continuity of management and business operations by providing emergency operating policies and procedures; description of who may declare a disaster leading to the implementation of the plan; assignment of powers, reporting structure, responsibilities and authority for emergency action; provisions for removal of emergency conditions; and the description of locations and powers of temporary main offices.

Sections 21 and 22 identify the individual members of the Disaster Management Executive Committee; assigns powers, authority and responsibilities to individual members; and provides for the termination of status. Section **21** is the appointment document, and Section **22** is the acknowledgement by each Committee member.

Section 23 is the resolution to be signed by the Board's Secretary upon the Board's approval, authorizing the Disaster Management Plan as the company's only plan, and authorizing the Disaster Management Team Chairpersons and Coordinators to carry out all provisions of the plan.

Section 24 is the annual re-certification resolution of the Disaster Management Plan based on the recommendation from the Disaster Management Team Coordinators and the Disaster Management Team Chairpersons.

Section 25 provides for the storage of resolutions and succession documentation, and access control codes and keys for facilities.

Section 26 contains listing of the confidential information retained in a separate binder, and stored with the Disaster Management Plan. This is not to be included in copies of the Disaster Management Plan distributed to individual offices and departments.

EXHIBIT 5–2

To: Office and Department Managers

From:

Date:

Re: Disaster Management Plan

Attached are three (3) items for your review and completion:

Draft recovery script: the information you supplied during the initial effort has been re-drafted into its present form. Please review, correct as necessary, and add any information requested in the [ ] or yellow-highlighted areas.

Emergency services and disaster information: this information will be entered into a master database to help your recovery efforts. Please research and complete each line as indicated.

Sample emergency staging area diagrams: Please draft similar diagrams for your office or facility, to be included in your plan and the master plan, and also to be used for training. The diagrams must contain:

Location of utility shutoff devices (gas, electric and water if appropriate); and

Directional pointers: N W E S

Access to this information may become critical during a disaster, both for your employees' well-being and for customer service. Please return all items to your Coordinators within ten (10) days. Thanks for your efforts, and please call me directly with any questions.

EXHIBIT 6

[This page intentionally left blank]

To:

From:

Date:

Re: Data Base Screens

To reduce our original research time and subsequent text entry, set up databases for all personnel, departments, offices and facilities, equipment, vendors and emergency service agencies.

This information is required for Sections 12.0 and 16.0 and the subsections dealing with equipment, supplies and furniture.

Then input the information once, print out each list at least annually, and update as a change occurs.

The company may already have most of this information in others forms; fixed asset list, employee list, supply/inventory list vendor (1099) list. If so, simply alter the database you have to include the information in these samples. The important thing here is to have current listings.

Quarterly updating is suggested.

EXHIBIT 7

[This page intentionally left blank]

SAMPLE PERSONNEL DATABASE: Last updated: [MM-DD-YYYY]

FIELD NAME	SAMPLE CONTENT
First name	[Fred]
Last name:	[Jones]
Title	[Customer Services Representative]
Home address	[7745 Bridgewater Lane, San Jose, CA 99999]
Home phone	[408/555 - 1212]
Work location/facility	[Capitol Office]
Work address	[312 E. Joshua St., San Jose, CA 99999]
Work phone	[408/555-1212]
Primary/Secondary team	[P]
Type of first aid card	[Advanced]
Date expires	[12/17/90]
Special skills	[CPR instructor, basic carpentry and electrical]

EXHIBIT 8

SAMPLE EQUIPMENT DATABASE: Last updated: [MM-DD-YY]

FIELD NAME	SAMPLE CONTENT
Category	[Personal computer]
Brand name	[IBM]
Model	[Model 30/OS/2]
Serial number	[IBM222881965T]
Storage location/facility	[Capitol Office]
Storage address	[312 E. Joshua St., San Jose, CA 99999]
Location phone	[408/555-1212]
Service provider	[Computer Repair Associates]
Service Provider phone	[408/555-1212]
Contact person	[Henry Aguilar]

EXHIBIT 9

SAMPLE FACILITIES DATABASE: Last updated: [MM-DD-YY]

FIELD NAME	SAMPLE CONTENT
Work location/facility	[Capitol Office]
Work address	[312 E. Joshua St., San Jose, CA 99999]
Work phone	[408/555-1212]
Number square feet	[1755]
Number electrical outlets	[47]
Number first aid kits	[4]
Number fire extinguishers	[6]
Emergency supplies stored for people	[28]
Special features	[Directly across the street from RV dealership, within one-half mile of motel and restaurant, within one mile of fire station and police station, within two miles from urgent care facility.]

EXHIBIT 10

SAMPLE EMERGENCY LIST DATABASE: Last updated: [MM-DD-YY]

FIELD NAME	SAMPLE CONTENT
Emergency agency	[National Guard Armory]
Address	[2895 Almaden Blvd., San Jose, CA 99999]
Emergency phone	[408/555-1212]
Business phone	[408/555-7777]
Emergency staging area	[National Guard Armory]
Address	[2895 Almaden Blvd., San Jose, CA]

Directions[Highway 280 to Allbright Ave. off-ramp. Take off- ramp east 2.7 miles to Almaden Blvd. Take Almaden Blvd. south for .6 miles, Armory on east side of the street.]

Contact person	[Freda Hanson]
To be contacted by	[Arlene Lopez, Ops. Officer]

Special information [Houses up to 575 people; food, shelter, water, medical and basic clothing needs. Parking for 400 vehicles; cellular communications link; emergency medical staging area for hospital overflow; command center for Fire Department.]

EXHIBIT 11

To: Office and Department Managers

From:

Date:

Re: Disaster Recovery Questionnaire

Attached is your disaster recovery questionnaire as described during the initial orientation meeting. Please complete the questionnaire and forward it to your Coordinators within ten (10) days. I've included an overview of the disaster management process with the questionnaire for reference.

All items requiring your completion are indicated by [ ], and necessary instructions or examples are included for each item.

Please remember that we are simply creating a direction and guidelines for ourselves and other employees to follow in the event of a disaster, not attempting to create an operations manual.

Keep in mind that, for whatever reason, you may not be the person who has to restore office or department operations, and that your employees may only have your written guidelines with which to work.

Locations of emergency staging areas and other standardized information will be added to your plan at the time of final printing, as will diagrams of the Command Centers.

In addition to reviewing this draft, please develop a list of your office's or department's forms or other special items (not routine office supplies or equipment) that should be stored for emergency use. Arrangements will be made to store those forms and special items with other emergency supplies to be stored at all office and department locations.

If you need help, please call me directly.

Thanks for your efforts

EXHIBIT 12

[This page intentionally left blank]

To: Office and Department Managers

From:

Date:

Re: Disaster Management Plan: Annual Review

Attached are three (3) items for your review, update research and completion:

Recovery script: Review, update and correct as necessary.

Emergency services and disaster recovery information: This information is maintained in a master database to help your recovery efforts. Review, update and correct as necessary.

Emergency staging area diagrams: Review, update and correct as necessary.

Access to this information may become critical during a disaster, both for your employees' well-being and for customer service.

Please return all items to me within ten (10) days.

In the next few weeks, you will receive copies of the updated information. Included with the new information will be your training handout, leader's guide and practical problem-solving exercise. Schedule your annual testing and training meeting, and advise me of the date when you return this material.

Thanks for your efforts, and please call me directly with any questions.

EXHIBIT 13

[This page intentionally left blank]

To:

From:

Date:

Re: Emergency supplies for offices and departments

Responding effectively to an emergency or disaster may require the use of various tools and materials. Review the following list and identify appropriate supplies and equipment, and develop a list and budget for the acquisition of the necessary items. This list should include all offices, departments and facilities.

EMERGENCY EQUIPMENT:

Battery (specify type) ________

Battery (specify type) ________

Battery (specify type) ________

Blanket, moon ________

Candle and matches ________

Camera/film (specify type film) ________

First aid handbook ________

Flashlight ________

Hammer ________

Hand axe ________

Knife, Swiss-army type ________

Garbage bag (33 gal/50 ct.) ________

Garbage bag (10 gal/100 ct.) ________

Gloves, latex (100 ct.) ________

Gloves, leather work ________

Pipe wrench ________

Pliers, channel lock ________

Putty knife, 1" ________

Pry-bar ________

Radio, AM-FM ________

EXHIBIT 14 - 1

Rope (100 ft. coil) ________
Screwdriver (Phillips) ________
Screwdriver (blade) ________
Shovel ________
Tape, electrical (100 ft.) ________
Tape, duct (150 ft.) ________
Tape, strapping (150 ft.) ________
Tape, yellow caution (100 ft.) ________
Vice-grips ________
Wire-cutters ________

EMERGENCY SUPPLIES:

First aid kit ________
Food (1 day supply) ________
Medical pack ________
Sanitary napkin ________
Water (1 day supply) ________

OFFICE SUPPLIES:

3-ring binder ________
Ball-point pen ________
Binder clip ________
Felt-tip pen ________
Floppy disk 5.25 ________
Floppy disk 3.5 ________
High-lighter pen ________
Label ________
Marking pen ________
Paper clips ________
Pencils ________
Pencil sharpener ________

EXHIBIT 14 - 2

Staple ________

Stapler ________

Staple remover ________

Tape, cellophane ________

PAPER:

Computer ________

Copy ________

Envelope, business ________

Envelope, manila ________

FAX ________

Index card ________

Post-It ________

Letterhead ________

Scratch pad ________

Second sheet ________

INTERNAL FORMS: (List internal company forms)

Company form ________

Company form ________

Company form ________

Company form ________

Company form ________

Company form ________

Company form ________

Company form ________

Company form ________

Company form ________

Company form ________

Company form ________

EXHIBIT 14 - 3

Company form ________
Company form ________
Company form ________
Company form ________
Company form ________
Company form ________
Company form ________
Company form ________
Company form ________
Company form ________
Company form ________
Company form ________
Company form ________
Company form ________
Company form ________
Company form ________
Company form ________
Company form ________

FORMS: (List other forms)

Form ________
Form ________
Form ________
Form ________
Form ________
Form ________
Form ________
Form ________

EXHIBIT 14 - 4

OFFICE EQUIPMENT:

COMPUTER: (each with appropriate specifications, versions, etc.)

Computer ________

Cable ________

Cable ________

Cable ________

Cable ________

Modem ________

Monitor ________

Printer ________

Printer ________

Software ________

Software ________

Software ________

Software ________

Software ________

Software ________

Software ________

Software ________

MISCELLANEOUS:

Adding machine ________

Calculator ________

Cash cart ________

Copy machine ________

Desktop basket ________

Fire safe ________

Light ________

Microfilm printer ________

EXHIBIT 14 - 5

Microfilm reader ________
Packing carton ________
Typewriter ________
Utility cart ________
Wastebasket ________
Write board ________

PHONES:

ACH ________
FAX ________
Modem ________
Number of lines ________
Number of phones ________
Special access ________

OFFICE FURNITURE:

DESKS:

Type ________
Type ________
Type ________

TABLES:

Conference ________
Utility ________
Printer ________

FILING CABINETS:

Bucket files ________
4 drawer letter ________
4 drawer legal ________
2 drawer letter ________
2 drawer letter ________
Printout style ________

EXHIBIT 14 - 6

CHAIRS:

Type: ________

Type: ________

Type: ________

Type: ________

OTHER:

After budget approval, assemble the following emergency supplies for each office and large department (small departments adjacent to other ones may share these items), and place all items in a 33 gallon plastic garbage can. Itemize the contents of the can on a separate piece of paper; tape the list to the outside top of the can, and seal the can with strapping tape.

Basic tools (hand axe with a hammerhead at one end, vice-grips, small pry-bar, small shovel, Swiss army-type knife, pipe wrench, channel lock pliers, 1" wide stiff putty knife, wire cutters, both blade and Phillips screwdrivers).

Two (2) pair of heavy-duty leather work gloves (size large), and one (1) box of 100 count latex surgical gloves.

One (1) battery-powered AM/FM radio, and extra batteries.

Two (2) flashlights and twice the number of batteries necessary to power each one. One (1) package of light sticks; several candles and matches.

Three (3) 50' rolls of duct tape; one (1) 50' roll of plastic electrical tape; and two (2) 25' rolls of strapping tape.

One (1) 100' coil of 1/4" nylon rope, and one (1) roll of yellow emergency "CAUTION" or "WARNING" marking ribbon.

One (1) box each of 100 count "tall kitchen" and 33 gallon plastic heavy-duty garbage bags with ties.

"Moon" or metalized emergency blankets for the total number of persons in the department or office.

Sanitary napkins and other similar absorbent materials.

Standard office supplies (paper, pens, pencils, stapler, paper clips and bankers clips, manila envelopes, heavy-duty rubber bands, etc.)

Department or office-specific forms or non-negotiable documents necessary to perform critical functions only.

"Instamatic"-type camera for insurance photos. If the camera uses a battery-flash, extra batteries; if it uses a "flashcube", extra cubes.

EXHIBIT 14-7

[This page intentionally left blank]

DISASTER RECOVERY

Sample Agreements

[This page intentionally left blank]

SAMPLE AGREEMENTS

Table of Contents*

Agreement **Page**

*The Sample Agreements Table of Contents references section page numbers (lower right corner).

[This page intentionally left blank]

Sample Services Agreement

Page 01

Agreement No. [COMPLETE]

!!!! THIS IS A SAMPLE FORMAT ONLY, AND IS NOT INTENDED TO BE A LEGAL DOCUMENT. ALTER IT TO SUIT YOUR NEEDS AND HAVE AN ATTORNEY REVIEW AND APPROVE THE FINAL AGREEMENT!!!!

This Agreement confirms the arrangements between [COMPLETE] (hereafter known as Backup Company) and [COMPLETE] (hereafter known as Lessee Company) for the disaster recovery services itemized as follows:

1.0 PURPOSE OF AGREEMENT

1.1 The purpose of this Agreement is to state the terms and conditions under which Backup Company will provide emergency backup facilities and services, as defined below, to the Lessee Company.

1.2 Using the Backup Company's personnel, facilities, supplies furniture, and equipment, Backup Company will provide Lessee Company with:

1.2.1 A designated area or areas in Backup Company's facility for storage of Lessee Company's records, forms and supplies. This area is described in Section 9.0 of this Agreement.

1.2.2 A designated area in Backup Company's lobby or other part of its facility that will be made available for use to Lessee Company after a Declaration of Disaster, to conduct the business of Lessee Company, including cash-handling functions if indicated. This area is described in Section 9.0 of this Agreement.

1.2.3 Furniture, equipment and supplies which will be made available to Lessee Company by Backup Company for Lessee Company's use after a Declaration of Disaster. This list is part of Section 9.0 of this Agreement.

1.2.4 A designated secure area or areas within Backup Company's vault for the storage of Lessee Company's negotiable and other dual custody documents.

1.3 Backup Company is responsible for the services listed in this agreement, and these services only. Backup Company assumes no responsibility for its inability to perform, due to the Lessee Company's failure to provide appropriate records, guidance and/or direction.

2.0 DEFINITION AND DECLARATION OF DISASTER

2.1 A disaster is any event, whether an act of God or otherwise, that will significantly affect any or all of the Lessee Company's operations.

2.2 A Declaration of Disaster is notification by an authorized representative of Lessee Company to an authorized representative of Backup Company that a Disaster has occurred and that Lessee Company is exercising its rights under this Agreement to use the designated facilities, equipment and supplies of Backup Company.

2.3 Authorized representatives of Backup Company are:

1. ______________ [COMPLETE] ______________________
2. ______________ [COMPLETE] ______________________
3. ______________ [COMPLETE] ______________________
4. ______________ [COMPLETE] ______________________

2.4 Authorized representatives of the Lessee Company are:

1. ______________ [COMPLETE] ______________________
2. ______________ [COMPLETE] ______________________
3. ______________ [COMPLETE] ______________________
4. ______________ [COMPLETE] ______________________

2.5 Either party may notify the other, in writing, at any time of a change in authorized representatives as indicated in Section 3.4 of this agreement.

3.0 TERM OF AGREEMENT AND CANCELLATION

3.1 The minimum initial term of this agreement is one (1) year; thereafter, automatic renewals are for a period of one (1) year, and shall continue in full force and effect thereafter until it is terminated by 30 days' written notice by either party to the other.

3.2 After a Declaration of Disaster by Lessee Company, Lessee Company shall be entitled to the use of the designated personnel, facilities, furniture, equipment and supplies of Backup Company listed in Section 9.0 of this Agreement for (7) seven days. This time period may be extended upon the mutual consent of the parties.

Backup Company shall make any arrangements which are reasonably necessary to assure Lessee Company has access to and the use of the designated personnel, facilities, furniture, equipment and supplies.

3.3 This Agreement includes the provision of data processing services by Backup Company to Lessee

Company: Yes _______; No _______ [COMPLETE]

(Both parties initial the appropriate box). If the "Yes" box is initialed, a Data Processing Service Agreement is attached to this Agreement and incorporated by this reference.

3.4 Any notice or correspondence required by this agreement shall be delivered personally, by certified United States Mail, or facsimile transmission as follows:

Correspondence to Backup Company: [COMPLETE]

Name: ____

Address: ______________________________________

C/S/Z: ____

Phone: ____________________ FAX: ________________

Correspondence to the Lessee Company: [COMPLETE]

Name: ____

Address: ______________________________________

C/S/Z: ____

Phone: ____________________ FAX: ________________

4.0 CONFIDENTIALITY

4.1 All statistical, financial and personnel data, as well as security codes and customer lists relating to the business of each party which are confidential, shall be kept in the strictest confidence by the other party and its employees, unless authorized for release by the affected party, or is compelled to be released by legal process. However, the foregoing obligation does not apply to any data, codes and customer lists which have become publicly available or that are rightfully obtained from third parties.

5.0 COMPENSATION

5.1 Lessee Company shall pay to Backup Company $__________ [COMPLETE] upon the signing of this Agreement and $__________ [COMPLETE] per month/year for the use of the designated storage area described in Section 9.0 of this Agreement.

5.2 Lessee Company shall pay to Backup Company $_________ [COMPLETE] per day for use of the designated facilities, furniture, equipment and supplies after the Declaration of Disaster.

5.3 Lessee Company shall also pay Backup Company the actual cost of telephone charges, fax charges, copy charges, charges for miscellaneous supplies not listed in Section 9.0 of this Agreement and other third-party charges incurred by Backup Company as a result of Lessee Company's use of the designated facilities, furniture, equipment and supplies after the Declaration of Disaster.

5.4 Further, Lessee Company shall compensate to Backup Company for the use of Backup Company's employees to perform work for Lessee Company after the Declaration of Disaster at the hourly rates set forth in Section 9.0 of this Agreement. Backup Company shall provide at least (1) officer to be present upon its premises at all times that this Agreement is activated.

6.0 EXCUSE FOR NONPERFORMANCE

6.1 Backup Company will be excused from performance under the terms of this Agreement if an authorized representative of Backup Company notifies an authorized representative of Lessee Company that Backup Company will not be able to perform because Backup Company's designated facilities have also been rendered unusable as a result of a Disaster.

6.2 "Force Majeure." None of the parties shall be deemed to be in default if performance of the obligations required by this agreement is delayed or becomes impossible because of any act of God, or earthquake, fire, strike, sickness, accident, civil commotion, epidemic, act of government, its agencies or officers, computer failure, or any legitimate cause beyond the control of the parties.

7.0 ATTORNEYS' FEES AND COSTS

If any action at law or inequity is necessary to enforce or interpret the terms of this Agreement, the prevailing party shall be entitled to reasonable attorneys' fees, costs and necessary disbursements in addition to any other relief to which such party may be entitled.

8.0 GOVERNING LAW, ENTIRETY OF AGREEMENT AND PARTIAL INVALIDITY

8.1 This Agreement shall be governed by the laws of the State of [COMPLETE]. It constitutes the entire Agreement between the parties regarding its subject matter. If any provision in this Agreement is held by any court to be invalid, void, or unenforceable, the remaining provisions shall nevertheless continue in full force.

8.2 Each company and its officers, employees, and/or agents are not liable to the other company, except for acts of willful misconduct or gross negligence by its officers, employees, and/or agents, in connection with the performance of this agreement. Each company agrees to defend, indemnify and hold the other company harmless from any loss, cost, expense, including attorneys' fees, that it suffers or incurs as a result of any claim or action made by a third party, arising out of this agreement.

9.0 SERVICES AND FEES

9.1 Personnel: Required Backup Company's personnel shall be compensated by Lessee Company at a rate equal to one and one-half (1 1/2) times the hourly rate of the position. Backup Company shall be responsible for the accurate calculation and payment of appropriate taxes and benefits to its employees. [COMPLETE: Describe other personnel needs here]

9.2 Facilities: Required use of Backup Company's facilities shall be calculated at a daily rate for: [COMPLETE]

Workspace	(@ _______ square feet)	$ ________
Parking	(@ _______ spaces)	$ ________
Secure vault	(@ _______ containers)	$ ________
Unsecured storage	(@ ______ square feet)	$ ________

[COMPLETE: Describe other facilities needs here]

9.3 Equipment: Required use of Backup Company's equipment shall be calculated at a daily rate for:

Computers (@ _______ per setup) $ _______________

(Includes workspace, printer, monitor, CPU, keyboard and necessary cables) [COMPLETE: Include any software or specialty items or configurations here]

Copy machines	(@ _______	machines)	$ __________
FAX machines	(@ _______	machines)	$ __________
FAX lines	(@ _______	lines)	$ __________
Microfilmer	(@ _______	per machine)	$ __________
Microfiche	(@ _______	per machine)	$ __________
Telephones	(@ _______	phones)	$ __________
Telephone lines	(@ _______	lines)	$ __________
Typewriters	(@ _______	machines)	$ __________
Vehicles	(@ _______	per vehicle)	$ __________

(Mileage @ .25 cpm)

[COMPLETE: Describe other equipment needs here]

9.4 **Furniture**: Required use of Backup Company's furniture shall be calculated at a daily rate for: [COMPLETE]

Chairs	(@ _______	chairs)	$ __________
Desks	(@ _______	desks)	$ __________
Tables	(@ _______	tables)	$ __________

[COMPLETE: Describe other furniture needs here]

9.5 **Supplies**: Required use of Backup Company's supplies shall be calculated at a daily rate for: [COMPLETE]

All general office supplies $ __________

[COMPLETE: Describe other supplies needs here]

9.6 **Cash and negotiable documents**: Required use of Backup Company's cash and negotiable documents shall be repaid in relation to actual usage, at a rate to be determined by

[COMPLETE: Describe formula here]

Cash limit available $ _________

Negotiable documents available _______ $ _________

Executed at ___[COMPLETE] ___, County of ___ [COMPLETE] ___,

State of ___ [COMPLETE] ___

on this ______ day of _____________, 20_____

"Backup Company"

By: ____________________________ Title: ______________________

Executed at ___ [COMPLETE] ____, County of ___ [COMPLETE] ___, state of ___

on this ______ day of _____________, 20_____.

"Lessee Company"

By: ____________________________ Title: _____________________

[This page intentionally left blank]

SAMPLE DATA PROCESSING SERVICES AGREEMENT

Page 01

Agreement No. ____________[COMPLETE]

!!!! THIS IS A SAMPLE FORMAT ONLY, AND IS NOT INTENDED TO BE A LEGAL DOCUMENT. ALTER IT TO SUIT YOUR NEEDS AND HAVE AN ATTORNEY REVIEW AND APPROVE THE FINAL AGREEMENT !!!!

This Agreement confirms the arrangements between ____ [COMPLETE] _____ (hereafter known as Backup Company) and _____[COMPLETE]_____ (hereafter known as Lessee Company) for the disaster recovery services itemized as follows:

1.0 PURPOSE OF AGREEMENT

1.1 The purpose of this Agreement is to state the terms and conditions under which Backup Company will provide emergency electronic data processing backup facilities and services, as defined below, to the Lessee Company.

1.2 Using the Backup Company's personnel, facilities, supplies furniture, and equipment, Backup Company will provide Lessee Company with:

1.2.1 A designated area or areas in Backup Company's facility for the processing of the Lessee Company's electronic data processing functions affected by the disaster. This area is described in Section 10.0 of this Agreement.

1.2.2 Furniture, equipment and supplies which will be made available to Lessee Company by Backup Company for Lessee Company's use after a Declaration of Disaster. This list is part of Section 10.0 of this Agreement.

1.2.3 A designated secure area or areas within Backup Company's vault for the storage of Lessee Company's electronic data processing media and records.

1.3 Backup Company is responsible for the services listed in this agreement, and these services only. Backup Company assumes no responsibility for its inability to perform, due to the Lessee Company's failure to provide appropriate records, guidance and/or direction.

2.0 DEFINITION AND DECLARATION OF DISASTER

2.1 A disaster is any event, whether an act of God or otherwise, that will significantly affect any or all of the Lessee Company's operations.

2.2 A Declaration of Disaster is notification by an authorized representative of Lessee Company to an authorized representative of Backup Company that a Disaster has occurred and that Lessee Company is exercising its rights under this Agreement to use the designated facilities, equipment and supplies of Backup Company.

2.3 Authorized representatives of Backup Company are:

1. ____________________ [COMPLETE] ____________________________

2. ____________________ [COMPLETE] ____________________________

3. ____________________ [COMPLETE] ____________________________

4. ____________________ [COMPLETE] ____________________________

2.4 Authorized representatives of the Lessee Company are:

1. ____________________ [COMPLETE] ____________________________

2. ____________________ [COMPLETE] ____________________________

3. ____________________ [COMPLETE] ____________________________

4. ____________________ [COMPLETE] ____________________________

2.5 Either party may notify the other, in writing, at any time of a change in authorized representatives as indicated in Section 3.4 of this agreement.

3.0 TERM OF AGREEMENT, CHANGES AND CANCELLATION

3.1 The minimum initial term of this agreement is one (1) year; thereafter, automatic renewals are for a period of one (1) year, and shall continue in full force and effect thereafter until it is terminated by 30 days' written notice by either party to the other.

3.2 After a Declaration of Disaster by Lessee Company, Lessee Company shall be entitled to the use of the designated personnel, facilities, furniture, equipment and supplies of Backup Company listed in Section 10.0 of this Agreement for (60) sixty days. This time period may be extended upon the mutual consent of the parties.

Backup Company shall make any arrangements which are reasonably necessary to assure Lessee Company has access to and the use of the designated personnel, facilities, furniture, equipment and supplies.

3.3 This Agreement includes the provision of reestablishing business operations by Backup Company to Lessee Company: Yes _______; No _______ [COMPLETE] (Both parties initial the appropriate box). If the "Yes" box is initialed, a Service Agreement is attached to this Agreement and incorporated by this reference.

3.4 Any notice or correspondence required by this agreement shall be delivered personally, by certified United States Mail, or facsimile transmission as follows:

Correspondence to Backup Company: [COMPLETE]

Name: ____

Address: ______________________________________

C/S/Z: ____

Phone: __________________ FAX: ____________________

Correspondence to the Lessee Company: [COMPLETE]

Name: ____

Address: ______________________________________

C/S/Z: ____

Phone: __________________ FAX: ____________________

4.0 CHANGES IN EQUIPMENT AND TESTING

4.1 Each company shall notify the other company, in writing, of any EDP equipment changes likely to affect either company's performance under this Agreement. Such notice shall be made as indicated in Section 3.4 of this Agreement, at least (90) ninety days prior to a change causing the configuration of its backup equipment, policies and procedures to be incompatible with the other company.

4.2. Backup Company shall make available to Lessee Company at least ____________ [COMPLETE] ____________ consecutive hours of testing time on its equipment each [MONTH/QUARTER/SIX MONTHS/YEAR], as a means of assuring that such equipment remains compatible for the purposes of this Agreement. The dates and hours to be made available for such testing will be mutually agreed upon by both companies, and each Company will agree upon (1) one week's prior notice for confirmation of the test(s).

4.3. Lessee Company may request from Backup Company an additional __________ [COMPLETE] ____________consecutive hours of testing time each [MONTH/QUARTER/SIX MONTHS/YEAR], and such additional time shall be made available by Backup Company within (1) one week of this request.

5.0 CONFIDENTIALITY AND SECURITY

5.1 All statistical, financial and personnel data, as well as security codes and customer lists relating to the business of each party which are confidential, shall be kept in the strictest confidence by the other party and its employees, unless authorized for release by the affected party, or is compelled to be released by legal process. However, the foregoing obligation does not apply to any data, codes and customer lists which have become publicly available or that are rightfully obtained from third parties.

5.2 Lessee Company agrees to utilize Backup Company's security procedures in effect at the time of any use of Backup Company's personnel, facilities, supplies furniture, and equipment.

6.0 COMPENSATION

6.1 Lessee Company shall pay to Backup Company $__________ [COMPLETE] upon the signing of this Agreement and $__________ [COMPLETE] per month/year for the use of the designated electronic data processing recovery area described in Section 10.0 of this Agreement.

6.2 Lessee Company shall pay Backup Company $_________ [COMPLETE] per day for use of the designated facilities, furniture, equipment and supplies after the Declaration of Disaster, in addition to reimbursing Backup Company for any repair or replacement costs incurred as a result of any physical damage to or disabling of its equipment, except for damage or disablement due to normal use.

6.3 Lessee Company shall pay Backup Company the actual cost of telephone charges, fax charges, copy charges, charges for miscellaneous supplies not listed in Section 10.0 of this Agreement and other third-party charges incurred by Backup Company as a result of Lessee Company's use of the designated facilities, furniture, equipment and supplies after the Declaration of Disaster.

6.4. Lessee Company shall compensate Backup Company for the use of Backup Company's employees to perform work for Lessee Company after the Declaration of Disaster at the hourly rates set forth in Section 10.0 of this Agreement. Backup Company shall provide at least (1) EDP Manager to be present upon its premises at all times that this Agreement is activated.

7.0 EXCUSE FOR NONPERFORMANCE

7.1 Backup Company will be excused from performance under the terms of this Agreement if an authorized representative of Backup Company notifies an authorized representative of Lessee Company that Backup Company will not be able to perform because Backup Company's designated facilities have also been rendered unusable as a result of a Disaster.

7.2 "Force Majeure." None of the parties shall be deemed to be in default if performance of the obligations required by this agreement is delayed or becomes impossible because of any act of God, or earthquake, fire, strike, sickness, accident, civil commotion, epidemic, act of government, its agencies or officers, computer failure, or any legitimate cause beyond the control of the parties.

8.0 ATTORNEYS' FEES AND COSTS

If any action at law or inequity is necessary to enforce or interpret the terms of this Agreement, the prevailing party shall be entitled to reasonable attorneys' fees, costs and necessary disbursements in addition to any other relief to which such party may be entitled.

9.0 GOVERNING LAW, ENTIRETY OF AGREEMENT AND PARTIAL INVALIDITY

9.1 This Agreement shall be governed by the laws of the State of [COMPLETE]. It constitutes the entire Agreement between the parties regarding its subject matter. If any provision in this Agreement is held by any court to be invalid, void, or unenforceable, the remaining provisions shall nevertheless continue in full force.

9.2 Each company and its officers, employees, and/or agents are not liable to the other company, except for acts of willful misconduct or gross negligence by its officers, employees, and/or agents, in connection with the performance of this agreement. Each company agrees to defend, indemnify and hold the other company harmless from any loss, cost, expense, including attorneys' fees, that it suffers or incurs as a result of any claim or action made by a third party, arising out of this agreement.

10.0 SERVICES AND FEES

10.1 Personnel: Required Backup Company's personnel shall be compensated by Lessee Company at a rate equal to one and one-half (1 1/2) times the hourly rate of the position. Backup Company shall be responsible for the accurate calculation and payment of appropriate taxes and benefits to its employees. [COMPLETE: Describe other personnel needs here]

10.2 Facilities: Required use of Backup Company's facilities shall be calculated at a daily rate for: [COMPLETE]

EDP recovery site $ __________

Workspace (@ _______ square feet) $ __________

Parking (@ _______ spaces) $ __________

Secure vault (@ _______ containers) $ __________

Unsecured storage (@ ______ square feet) $ __________

[COMPLETE: Describe other facility needs here]

10.3 **Equipment**: Required use of Backup Company's equipment shall be calculated at a daily rate for: [COMPLETE]

Computers (@ _______ per setup) $ _________

(Includes workspace, printer, monitor, CPU, keyboard and necessary cables) [COMPLETE: Include any software or specialty items or configurations here]

Copy machines (@ _______ machines) $ _________

FAX machines (@ _______ machines) $ _________

FAX lines (@ _______ lines) $ _________

Microfilmer (@ _______ per machine) $ _________

Microfiche (@ _______ per machine) $ _________

Telephones (@ _______ phones) $ _________

Telephone lines (@ _______ lines) $ _________

Typewriters (@ _______ machines) $ _________

Vehicles (@ _______ per vehicle) $ _________

(Mileage @ .25 cpm)

[COMPLETE: Describe other equipment needs here]

10.4 Furniture: Required use of Backup Company's furniture shall be calculated at a daily rate for: [COMPLETE]

Chairs (@ _______ chairs) $ _________

Desks (@ _______ desks) $ _________

Tables (@ _______ tables) $ _________

[COMPLETE: Describe other furniture needs here]

10.5 Supplies: Required use of Backup Company's supplies shall be calculated at a daily rate for: [COMPLETE]

All general office supplies $ __________

[COMPLETE: Describe other supply needs here]

Executed at [COMPLETE], County of [COMPLETE], State of [COMPLETE] on this ______ day of ____________, 19_____.

"Backup Company"

By: ___________________________ Title: _____________________

Executed at [COMPLETE], County of [COMPLETE], State of [COMPLETE]
on this ______ day of ____________, 19_____.

"Lessee Company"

By: ___________________________ Title: ____________________

[This page intentionally left blank]

DISASTER RECOVERY

Sample Forms

[This page intentionally left blank]

SAMPLE FORMS

Table of Contents*

*The Sample Forms Table of Contents references section page numbers (lower right corner).

[This page intentionally left blank]

DISASTER RECOVERY LOG

DATE: __________ TIME STARTED: ________ TIME ENDED: __________

OFFICE/DEPT: ____

OPERATOR: ____

OFFICE/DEPT CONTACT: ____

PHONE: ( )/_______________ FAX: ( )/________________

LOCATION: ____

ALTERNATE: ____

PHONE: ( )/______________ FAX: ( )/________________

LOCATION: ____

RECOVERY SITE: ____

RECOVERY SITE: ____

RECOVERY SITE: ____

Notes ____

____

____

____

____

____

____

EXHIBIT 1

[This page intentionally left blank]

CHRONOLOGICAL LOG OF EVENTS

DATE: ____________ TIME STARTED: _______ TIME ENDED: _______

LOCATION: ____

MANAGER IN CHARGE: ________________ OPERATOR: _____________

Time	Person	Activity/request	Disposition	Time

EXHIBIT 2

[This page intentionally left blank]

TESTING OF SOLUTIONS DOCUMENTATION

On ______________________ the Disaster Management Plan solutions to the scenario indicated below were tested with the following results:

SCENARIO:

____

____

____

____

____

____

Test conducted by: ____

Location/Department: ____

Date: _________________________ Time: ______________________

Results:

____

____

____

____

____

____

____

____

____

____

____

EXHIBIT 3

[This page intentionally left blank]

RECOVERY WORKSHEET for:

____

Following the onset of an emergency or disaster requiring the activation of the company's Disaster Management Plan, the following procedures must be implemented to recover/stabilize operations:

1. Follow emergency instructions for specific position descriptions, addressing employee and customer safety first.
2. Assess any injury to employees and customers, and damage to facilities.
3. Temporarily close and secure the facility, if necessary.
4. Contact appropriate emergency services, if necessary.
5. Begin documenting the effects of the disaster and actions taken; secure all assets and records.
6. First attempt to contact the Disaster Management Team Coordinators with a damage assessment and actions taken report, and act upon instructions received. If the Coordinators are unavailable, contact Chairpersons for instructions.
7. If all documented attempts to communicate with Coordinators and Chairpersons have failed, the Department/Office Leader or designate is authorized to initiate reasonable and prudent responses necessary to minimize potential:
    1. Injuries to personnel and customers;
    2. Damage to facilities; and
    3. Loss of assets and records.

Additional requirements

EXHIBIT 4

[This page intentionally left blank]

RESTORATION WORKSHEET

Following the recovery/stabilization of operations, the following procedures must be implemented to restore minimum levels of service:

1. Based upon the actual results of the disaster, review and temporarily revise the disaster response plan appropriately.
2. Assign the performance of critical tasks to designated employees, and initiate and maintain reporting structure and requirements.
3. Conduct a damage assessment of all equipment and supplies. Arrange for the transportation of replacement equipment and supplies, if appropriate.
4. Verify that all necessary equipment and supplies are available to perform critical tasks.
5. Retrieve data files, vital records, and other critical work.
6. Provide for physical site security, and institute other security policies and procedures.
7. Clear the work areas and provide temporary repairs as appropriate.
8. Continue documenting the effects of the disaster, and all actions taken to restore services. Photograph the impact areas if possible.
9.
10.
11.
12.

EXHIBIT 5

[This page intentionally left blank]

RESUMPTION WORKSHEET For:

____

Following the restoration of basic operations, the following procedures must be implemented to resume normal levels of service:

1. Verify that basic operations have been fully restored, and notify the Disaster Management Team Coordinators of the completed restoration.

2. Determine personnel, equipment and supply needs to complete the restoration of normal services.

3. Replace or restore equipment as appropriate.

4. Continue documenting the effects of the disaster, and all actions taken to restore services.

5.
6.
7.
8.
9.
10.
11.
12.

EXHIBIT 6

[This page intentionally left blank]

RECONSTRUCTION WORKSHEET FOR:

____

Following the resumption of normal operations, the following procedures must be implemented to reconstruct this department to the level of operation existing prior to the emergency:

1. ____
2. ____
3. ____
4. ____
5. ____
6. ____
7. ____
8. ____
9. ____
10. ____

EXHIBIT 7

[This page intentionally left blank]

EMPLOYEE PERSONAL PROFILE

The information requested in this questionnaire is voluntary and confidential, and is not to be used for any purpose other than during an actual emergency. The contents of this questionnaire shall be kept in a sealed envelope in a secure area, and shall not be opened unless in the case of an actual emergency. The contents of this questionnaire and a photograph shall be updated annually during your performance evaluation.

PERSONAL IDENTIFYING INFORMATION:

Employee: ______________________________

Nicknames or other names used: ______________________________

Employment classification: ______________________________

Employment location: ______________________________

Permanent residence: ______________________________

______________________________ Phone: ______________

Secondary residence: ______________________________

______________________________ Phone: ______________

Other employment, if applicable: ______________________________

Date of birth: / / / Place of birth: ______________________________

Name of hospital: ______________ Mother's name: ______________

Race: ______ Sex: ______ Complexion: ______________

Height: _____ Weight: ____ Hair color: ________ Eye color: ________

Scars/marks/tattoos: ______________________________

Hobbies: ______________________________

Are your fingerprints on file with the company? ____________

Is a current photograph on file with the company? ___________

FAMILY AND EMERGENCY NOTIFICATION INFORMATION:

Marital status: ______________ Anniversary date: __________

Name of spouse/roommate: ______________ Nickname: ____________

EXHIBIT 8-1

Number of children: _______ Children's names are:

____ Birth date: ______________
____ Birth date: ______________
____ Birth date: ______________
____ Birth date: ______________

PERSONS TO CONTACT IN CASE OF EMERGENCY:

Name: ____________________________ Phone: ________________
Address: ________________________________ Relationship: ______________
Name: ____________________________ Phone: ________________
Address: __________________________ Relationship: ____________
Name: ________________________________ Phone: ___________________
Address: __________________________ Relationship: ____________

IMMEDIATE CLOSE RELATIVES:

Name: ____________________________ Relationship: ____________
Address: __________________________ Phone: _______________
Name: ____________________________ Relationship: ___________
Address: __________________________ Phone: _______________
Name: ____________________________ Relationship: ___________
Address: __________________________ Phone: ________________

OTHER PERSONS LIVING OR WORKING IN HOUSEHOLD:

Name: ________________________ Relationship: ___________
Address: ______________________ Phone: _______________
Name: ________________________ Relationship: ___________
Address: ______________________ Phone: _______________

MOTOR VEHICLES USED BY FAMILY

Year: ___ Make: _______ Model: _________ Color: __________
License: ______________ Driven by: ______________________
Year: ___ Make: _______ Model: _________ Color: __________
License: ______________ Driven by: ________________________

EXHIBIT 8-2

MEDICAL INFORMATION:

Blood type: _____ Allergic to: ______________________________________

Medical condition requiring treatment or medication:

____

____

Medication required: _________________ Dosage: ____________________

Personal physician: __________________ Phone: _____________________

Address: ____

AUTHORIZATION FOR THIS PHYSICIAN TO RELEASE CONFIDENTIAL INFORMATION IN THE EVENT OF AN EMERGENCY SITUATION REQUIRING TREATMENT:

Signature: _________________________________ Date: _____________________

EMPLOYEE:

After completing this form, place the form into the envelope provided and seal the envelope. Write your name (last, first, middle initial) and Social Security number on the face of the envelope. Deliver the sealed envelope to your reviewer.

Please include any other information you feel is necessary on a separate page, to be included with this profile.

REVIEWER:

Ensure that the employee's envelope is securely sealed and completed as indicated. Initial and date the face of the envelope, and store in a secure area.

EXHIBIT 8-3

[This page intentionally left blank]

SAMPLE FACILITY DIAGRAMS

Develop and save diagrams of Emergency Exits for each work area, floor and building

EXHIBIT 9

[This page intentionally left blank]

DISASTER RECOVERY

Training Handout

[This page intentionally left blank]

TRAINING HANDOUT

Table of Contents*

*The Training Handout Table of Contents references section page numbers (lower right corner).

[This page intentionally left blank]

1.0 INTRODUCTION TO DISASTER MANAGEMENT AND BUSINESS RECOVERY

A disaster is any event that will significantly impact the company's operations. Anything traditionally, this means fire, flood, and earthquake. Non-traditionally, this means a toxic waste dispersion, computer system crash, power outage, labor strike, community illness and many others.

No one is exempt from a disaster, and most people are unprepared if a disaster erupts. We should prepare as if the disaster will exceed our ability to respond effectively: a worst-case scenario.

Disasters result from three (3) types of incidents, caused by:

1. **Natural or cataclysmic events** (i.e., earthquake, fire, flood, storm);
2. **Human behavior** (i.e., robbery, bomb threat, arson, hostage event, transportation strike); and
3. **Technological breakdowns** (i.e., power outage, computer crash).

The key to disaster management is to have a plan in place when the disaster strikes. A set of simple, effective guidelines and procedures for all people to follow. Just as a ship without a rudder is at the mercy of the tides, a company without a plan is at the mercy of events.

Human beings tend to make inappropriate decisions during a crisis. If we have a plan for coping with most emergencies already prepared and shared with others, we stand a better chance of surviving any emergency and recovering rapidly.

The emergency we prepare for is one that will cause one or more of the following results to occur in your community:

- The governor of the state or another competent authority declares the community a disaster area.
- Martial law is declared, the National Guard is activated, and basic civil rights are suspended.
- A state of total lawlessness in the community; law enforcement agencies are unable to cope with the results, even with the assistance of the National Guard.
- One third of the community dies immediately, or becomes so gravely disabled that they are unable to care for themselves or others.
- One third of the community is unavailable, unable or unwilling to help in disaster recovery efforts.
- One third of the community becomes responsible for the welfare of the remainder of the community, assuming roles and responsibilities for which they may be untrained and unprepared to handle.

- Traditional roles and routine activities within the community are temporarily suspended. This causes the reassignment of personnel to fulfill necessary roles and functions, with temporary powers, duties and responsibilities.
- **A 72-hour interruption of**:
  - **Basic emergency and transportation services**, such as police, fire and medical assistance, and both public and private forms of transportation. Routine calls for service are not answered, and the agencies' priorities will involve life-threatening situations only. Food, water and survival supplies. Deliveries to food stores and hospitals are interrupted or hijacked. People immediately strip existing supplies of all kinds, stockpiling their own reserves. Water pipes break, crack or clog, reducing or eliminating water flow.
  - **Electrical power, sanitation services, sewer.** Power lines are destroyed, overloaded or collapse. Water flow is restricted, and sewer pipes break, crack or clog and back up. Effluent reservoirs burst and flood the surrounding area with untreated waste.
  - **Vendors of products and services** will be unable to fulfill their roles, either because the geographical area affected by the disaster is closed to outside traffic; phone lines are down and information cannot be shared; or the vendors will be too busy answering calls for service from other clients also affected by the disaster.

We plan for a disaster as if we will not have any outside help for at least three to five days. We should be self-sufficient during that minimum amount of time, and preferably longer.

2.0 GOALS OF DISASTER MANAGEMENT AND BUSINESS RECOVERY

The three primary goals of disaster management and business recovery planning are to:

1. Eliminate or reduce the potential for injuries or the loss of human life, damage to facilities, and loss of assets and records. This requires a comprehensive assessment of each department within the company, to insure that appropriate steps have been taken to:
   - Minimize disruptions of services to the company and its customers;
   - Minimize financial loss;
   - Provide for a timely resumption of operations in the event of a disaster; and
   - Reduce or limit exposure to potential liability claims filed against the company, and its directors, officers and other personnel.

2. Immediately invoke the emergency provisions of Disaster Management Plan to stabilize the effects of the disaster, allowing for appropriate assessment and the beginning of recovery efforts. We then contain the disaster to its least impact and provide for the fastest recovery.
3. Implement the procedures contained in the Disaster Management Plan according to the type and impact of the disaster. When we implement these procedures, we must prioritize all recovery efforts as follows:
   - **Employees**: Not only must we help to ensure their survival as a basic human concern, but for their performance in helping other persons on company premises when the disaster strikes.
   - **Customers**: As we do with employees, we must help to ensure the survival or care of customers affected by the disaster; physically, mentally, emotionally and financially.
   - **Facilities**: After ensuring the safety of employees and customers, we then secure the facility as shelter for both people and assets.
   - **Assets**: Although for the most part recoverable, assets will only be addressed after people and facilities are secure.
   - **Records**: As with assets, we will only address the recovery and reconstruction of important records when all people are cared for, facilities secured, and all assets have been audited and stored.

3.0 THE DISASTER MANAGEMENT PLAN

To be effective, a disaster management and business recovery plan contains:

- The company's philosophy, mission statement and goals regarding disaster management and business recovery.
- Written and approved executive succession instructions.
- The appointment of a temporary Disaster Management Executive Committee for the term of the emergency, who may also act in the absence of the company's Board of Directors.
- Clearly defined guidelines and scope of all disaster management and business recovery efforts, based upon a thorough risk-assessment exercise.
- Clearly defined duties, authority and responsibilities for each employee classification, with designated primary and alternate department leaders and staff personnel to manage critical functions.
- A business recovery plan (operations manual) for each office, department, facility and function within the company, and for essential service vendors.
- Designated and equipped sites for the assembly of personnel for each phase of the disaster management and business recovery effort.

- A well-documented testing and evaluation process conducted at specified intervals, and at least annually.
- A comprehensive training program for all personnel at all facilities, conducted at specified intervals, and at least annually. This includes the identification and operation of utility shut-off devices; location of emergency staging areas; basic first aid and survival techniques as appropriate; and emergency responsibilities and re-assignment plans for positions as appropriate.
- Written copies of the final Disaster Management Plan distributed to office and department leaders, including a complete list of all emergency response agencies and facilities.

4.0 THE DISASTER MANAGEMENT TEAM

The company's Disaster Management Team is drawn from the company's available personnel resources, and may include service vendors. The Team functions as a strategic planning and tactical response unit. Team members may be assigned temporary duties, responsibilities and a level of authority beyond their normal employment classification.

■ DISASTER MANAGEMENT TEAM ■

Organizational Chart:

Executive Committee:

Policies and Procedures:

Organization

Chairpersons:

Leadership and Administration:

Organization

____

____

Coordinators:

Management and Implementation: Organization

____

____

Department/Branch Leaders:

Management and Implementation: Department, Branch or Function

____

____

Staff:

Implementation: Individual Assignment

The **Disaster Management Team** is comprised of executives, managers and department heads with functional responsibilities involving several offices, departments or facilities. The Team may include sub-groups as appropriate.

The **Disaster Management Executive Committee** is charged with the responsibility for creating a comprehensive Disaster Management Plan for the company. The Committee is composed of the company's President/CEO, selected executives and senior managers, and may contain members of the company's Board of Directors. The Disaster Management Plan is used as the guiding mechanism for fulfilling the company's duties and responsibilities regarding local, state and federal regulations.

The company's **Disaster Management Executive Committee** members are, by name:

____

____

____

____

____

____

The **Disaster Management Team Chairpersons** are responsible for assigning the research, development and implementation of the Disaster Management Plan throughout the organization.

The Chairpersons are also responsible for assuming control over business operations during and after the disaster, until business operations return to normal. The Chairpersons are responsible for providing effective leadership and administration of the company's recovery efforts.

The company's Disaster Management Team Chairpersons are, by name:

____

____

____

____

The **Disaster Management Team Coordinators** are charged with the responsibility for creating a comprehensive Disaster Management Plan for the company, and for implementing the decisions made by the Chairpersons and the Disaster Management Executive Committee during a disaster. The Coordinators are responsible for providing effecting management and implementation of the company's recovery efforts.

The company's Disaster Management Team Coordinators are, by name:

Coordinator # 1 __________________________________

Coordinator # 2 __________________________________

Coordinator # 3 __________________________________

Coordinator # 4 __________________________________

The Disaster Management Team Department/Office Leaders and service vendors are responsible for acting as Coordinators for their respective areas of responsibility, and report directly to the Disaster Management Team Coordinators.

The company's Disaster Management Department/Office Leaders are indicated on the attached list.

Each employee is responsible for assisting management, as appropriate, in the following activities:

- Immediately securing all cash and negotiable documents, locking cash and dual custody containers, and keeping the key.
- Checking the immediate area for hazards created by the disaster.
- Determining if anyone needs medical assistance, and assisting if possible.
- Immediately reporting to supervisors for instructions; advising the supervisor of potential hazards, persons in need of medical or other assistance, or other pertinent information.
- Following instructions and performing tasks as assigned by any supervisor.
- Evacuating other personnel and customers from the facility, if necessary.

5.0 EMERGENCY STAGING AREAS AND FACILITY EVACUATION

Emergency staging areas are designated assembly sites to be used in the event of an evacuation of the facility. Each employee is responsible for assisting in a simple, safe, and rapid evacuation from the facility, based upon the following priorities:

- **PEOPLE**: Employees and customers. If the danger is outside the facility, stay inside: if the danger is inside the facility, go outside. Pay attention to fallen building materials, downed electrical wires, shattered glass, traffic and other hazards.
- **PLACES**: Secure the facility before evacuating, if appropriate. Take all keys, close vaults, and pay attention to hazards created by the disaster. Shut off utility devices, if appropriate.
- **THINGS**: Secure and protect records and assets before evacuating, if appropriate. Take all keys, close vaults, and pay attention to hazards created by the disaster.

The Department/Office Leaders are responsible for:

- Assigning at least two (2) employees as a "roundup" team, responsible for physically searching the facility for employees and customers, and moving these persons to the designated evacuation door.
- Assigning at least two (2) employees as the "door monitor" team, responsible for counting all persons leaving the facility, and for determining the need for medical assistance.
- Assigning at least two (2) employees as the "escort" team, one employee responsible for leading the evacuation in an orderly manner, single-file: the other

employee responsible for monitoring progress and gathering stragglers back into line.

When all persons have arrived at the emergency staging area:

- The "door monitor" team becomes responsible for recounting all persons, and advising the Office/Department Leader of the results and the need for any medical attention.
- The "roundup" team becomes responsible for rendering first aid or emotional support.
- The "escort" team becomes responsible for keeping all persons assembled at the site until emergency services agencies arrive.

The company's designated emergency staging areas are designated on diagrams, and are and should be included at the end of this section and described as:

- Primary Staging Area
- Secondary Staging Area

6.0 TELECOMMUNICATIONS

Telephone links will likely be among the first casualties of a community-type disaster. Telephone contact is imperative for both personal and business survival, and the lack of current information regarding conditions may lead to inappropriate decisions.

During a disaster that affects telephone communications, telephone links may still be in use in another area code, or by public phone. The company maintains a list of public telephone numbers and locations for each facility, and these phones are among the first to be restored.

The company has also designated several telephone numbers or extensions for emergency use. These numbers are:

- Hotline number(s) for use by employees only, for instructions or assignments.
- Third-party relay number(s) outside of this area code:
- Disaster Management Team Chairpersons, by name:
- Disaster Management Team Coordinators, by name:

7.0 ALTERNATE WORK SITES

A community-type disaster will affect telephone lines, streets and highways, and accessibility to buildings. Alternate work sites have been established for disaster recovery efforts. The company's alternate sites are:

Primary: ______________________________

Address: ______________________________

City: ______________________________

State: ______ ZIP: ______________________

Phone: ______________________

FAX: ______________________

Secondary: ______________________________

Address: ______________________________

City: ______________________________

State: ______ ZIP: ______________________

Phone: ______________________

FAX: ______________________

8.0 MEDIA

It is natural that any disaster will attract media attention from television, radio and newspapers. Media representatives have a legal right to gather information for publication. Employees have the same right to refuse to be interviewed.

The company authorizes only certain designated persons to provide information to the media, and specifically prohibits all other employees from media interviews. All employees not so designated are to courteously refer all requests for media interviews to one of the following people:

- President, CEO or designate;
- Disaster Management Team Chairpersons; or
- Disaster Management Team Coordinators.

9.0 PERSONAL VS. BUSINESS RESPONSIBILITIES

The company recognizes that, once the initial shock of surviving a disaster has subsided, employees immediately become concerned with personal issues. It's a natural and normal reaction.

It is the company's belief that employees must resolve personal issues first, and only after those issues have been successfully resolved will employees be mentally, physically and emotionally prepared to assist in the company's disaster recovery efforts.

The company suggests that all employees follow these priorities before attempting to handle their responsibilities in the company's disaster management and business recovery efforts, if a disaster occurs during either working or non-working hours:

Contact the company's emergency telephone number, and advise the company of your location and emergency phone number, safety, issues to be resolved, estimated time to resolve issues, and other appropriate information.

Locate and determine the condition of spouses, children, relatives, roommates, other appropriate persons, and pets.

If local telephone communication links are unavailable, establish a third-party relay with a friend or relative whose area code is in another geographical region.

Determine the condition of residences, roadways and utilities affecting immediate transportation, living conditions, or property loss. Report to an assigned work location only when it is safe to do so, and only when you are prepared to participate in the company's disaster management and business recovery efforts.

The company also encourages all employees to carry emergency supplies in their vehicles, in the event that they are stranded between locations. Consider keeping supplies similar to those listed below in the trunk of your vehicle(s):

- Basic tools, including a hand axe with a hammerhead at one end, pliers, small pry-bar, small shovel, Swiss army-type knife, can opener, fire extinguisher.
- A pair of heavy-duty leather work gloves (size large); a battery-powered AM/FM radio, and extra batteries; a flashlight with twice the number of batteries necessary to power it; and matches.
- A roll of duct or strapping tape; a 100' coil of 1/4" nylon rope; an emergency blanket and warm clothing, including comfortable shoes; a first aid kit and first aid instructions; sanitary napkins and other similar absorbent materials, and a roll of paper towels; plastic garbage bags (33 gallon).
- Food, water and special medical supplies sufficient to supply you for one (1) day or more. This includes eyeglasses, special foods, eating utensils, drinking container, prescriptions, infant formula and clothing if appropriate, and related personal items.
- Writing materials, and a list of family members' and other important phone numbers, including a friend or relative in another state that you and your family can call to relay safety information.
- Anything else you believe is important; then stock these same supplies at home. Advise family members of the location and use of these supplies, and appropriate equipment.

10.0 ATTACHMENTS

1. Disaster Management Department/Office Leaders list.
2. Facility diagrams with emergency exits.
3. Directions to alternate business operation sites.

VIOLENCE IN THE WORKPLACE 430

[This page intentionally left blank]

VIOLENCE IN THE WORKPLACE

TABLE OF CONTENTS*

Segments **page**

* The Violence in the Workplace Table of Contents references section page numbers (lower right corner).

FOREWORD

STRATEGIES & TACTICS FOR WORK & HOME

Assault is the most feared personal crime.
You can be assaulted physically, mentally and emotionally. Statistically, one in every four Americans will become the victim of a violent act during your lifetime, either at home, at work — or any place in between. Assaults do not just happen to "other people". Anyone — even you — can become a victim.

The **Coping with Crimes of Violence Manual** uses a simple, straightforward and logical approach and problem-solving scenarios to help you understand the concept of personal protection — both at work and at home. By understanding the cause and effect relationships between personal and work habits, a company's policies and operating procedures and the behaviors exhibited by both victims and offenders, you can implement both a comprehensive personal and company-wide security and safety program that will significantly reduce most safety risks.

The Coping with Crimes of Violence Manual is suitable for your personal use and for all types of businesses — including retail, wholesale, service and financial services industries. This information is essential for preventing crimes of violence and for surviving violent events if your prevention efforts fail. Victims who prevent or survive violent events generally report having had a defensive plan of action, consisting of a strategy and thoughtful tactics and response techniques.

The Coping with Crimes of Violence Manual is designed especially for those people who are responsible for any company's safety and soundness: directors, executives, compliance officers, security directors, auditors and operations managers. The Manual is a strategic planning, training and reference tool for helping you to decide what to do before, during and after a violent event.

Previous crisis or emergency management experience is helpful but it is not necessary for you to understand and to use the information. You will benefit if you are responsible for or assist with the development of safety strategy and emergency response plans, conducting operational reviews, training employees or writing policy and procedures.

The Coping with Crimes of Violence Manual is about options — not guarantees. Tactics that work for one person may not work for another. The information contained in the Manual has been acquired from thousands of interviews conducted during law enforcement field exercises; many more years of research; and the sharing of other professionals' experiences in conversations, seminars and workshops. And these tactics could be all wrong — for you. You may have used a better way to resolve an event, or you may be physically, mentally, spiritually or emotionally incapable of using these tactics.

Some of the suggestions are repulsive to "normal" people; many challenge your courage; all are designed to accomplish a common goal — to prevent you from becoming a victim of a violent act, or at least to reduce the likelihood of serious injury. The choices — and your success or failure — are all yours. You can, after all, do everything right and still lose — and everything wrong and still win.

You have more work to do if you are truly committed to developing a safe working environment for your company, including:

- Conducting a survey of your employees to help you determine if, in fact, you even have a potential problem. Your insurance company's risk management department may provide the survey form — or even conduct the survey for you.
- Contacting your local, state and federal emergency and crisis management agencies, insurance providers and other community resources for more information. The agencies, providers and other service groups often offer literature, signage, training programs and counseling referrals that you can adapt to your workplace.
- Using a computer online service to locate and contact other resources, such as articles in periodicals, online focus groups and other similar companies that have experienced the same events that you are trying to prevent. You can benefit from other peoples' mistakes.

The process of developing and maintaining safe places to live and work is an evolutionary one. Learn to be conscious, ever-vigilant and adaptable to life's circumstances. As you convert this information to fit your personal and work lives, remember that coping with crimes of violence involves four components: planning, prevention, response and recovery. Inattention to one component may jeopardize the success of the other three.

The Coping with Crimes of Violence Manual is dedicated to the memory of William B. Skelton, whose tireless efforts in promoting personal safety strategies and tactics for the students, business persons and law enforcement officers of California did not go unnoticed.

1.0 THE PROBLEM

1.1 OVERVIEW

Violence in the workplace has tripled in the last ten years, while the nation's population has increased by less than 10%. Ten years ago, we rarely heard about a violent situation in the workplace. If we did, we did not think of it as a commonplace happening — it was a singular, isolated and very shocking event.

The first workplace violence incident to capture the world's attention happened in 1966. Charles J. Whitman, who had been a mild-mannered Eagle Scout, climbed up the clock tower at the University of Texas in Austin. He carried a sniper rifle, a variety of other weapons, and hundreds of rounds of ammunition. After he barricaded himself in the tower, he began systematically to murder everyone within killing distance. Before the police gunned him down, he had murdered thirteen people and wounded another thirty-one — a staggering total of forty-four casualties in and around the campus

Today, we hear about workplace violence incidents on almost a daily basis. Those incidents may be more or less dramatic and they are far more frequent. Almost every time we read the newspaper, listen to the radio or watch TV, we learn about incidents like these:

- An employee arrives for work too early and interrupts a burglary, prompting a hostage event;
- A manager is kidnapped, held for ransom and killed after the money is paid;
- An ex-employee returns to the workplace and shoots his former supervisor — and many other people who have no direct connection to the ex-employee's grievance;
- An employee's domestic partner or other family member stalks the employee, and assaults him/her in the company's parking garage;
- A customer is stabbed during an ATM or night deposit robbery;
- A female courier stops at an intersection, is dragged from her vehicle and shot during a carjacking; and
- A group of gang members stage a takeover robbery, execute several employees and customers, and then finish the robbery.

Workplace violence has become one of the most significant occupational — and financial — hazards that employers, supervisors and employees are likely to face. Managing the threat of workplace violence and responding to potentially violent incidents in the workplace are among the greatest concerns faced by security directors, owners, directors, officers and managers. A legal settlement and related costs regarding a single serious incident can cost a company millions of dollars.

A company can never create a totally safe work environment, while simultaneously employing people and providing service to customers. A company can, however, devise strategies and tactics to reduce risk. To create such a safety program, the whole issue of workplace violence must be logically and critically examined. We have to know who does it, what they do, how they do it, and why.

1.2 ASSAULT

Acts of workplace violence generally begin with simple assault. Assault is the most feared personal crime. Assaults may be committed physically, mentally or emotionally. Statistically, one in every four Americans will be the victim of a violent act during their lifetimes, either at home, at work — or any place in between. The likelihood of becoming a victim of a physical assault is present in incidents involving a (n):

- Burglary (e.g., residential, commercial, vehicle).
- Robbery (e.g., holdup, takeover, ATM, carjacking).
- Extortion (e.g., self, family, employee, hostage).
- Relationship (e.g., spouse, children, neighbor, coworker, customer, vendor).
- Terrorist act (e.g., bombing, arson, vandalism, sabotage, assassination).
- Random act (e.g., panhandling, pick-pocketing, mugging, sexual assault).

1.3 VICTIMS

A victim is a person who is hurt — a casualty of an event. We often and unknowingly create opportunities for events to occur which make us become victims. We become victims because we make one or more critical errors in judgment (mistakes) that then cause us to make inappropriate decisions. We make these mistakes because we are:

- Ignorant: we have no knowledge (education) about this type of event.
- Naive: we have no experience (training) about this type of event.
- Stupid: we know the "right" thing to do — but we do something else instead.

We make these kinds of mistakes all too frequently. We walk to our cars in dimly-lit parking lots without examining our surroundings. We leave our car doors unlocked. We open our wallets to strangers on the street corner who ask for money, or turn our backs to these people after refusing to give them money. We give personal information to strangers who call us — even when the callers fail to identify themselves appropriately. We remain in the presence of people who have been drinking excessively and are becoming aggressive. We try to intervene when people are arguing.

The profile of the usual victim and the description of the typical crime that results in workplace homicide are surprising. Most workplace homicide victims are white males between the ages of twenty-five to thirty-four. Most homicides occur during a robbery. Most offenders use a firearm. Most victims did not know their attackers. Most victims worked in retailing or other service industries — exchanging money and delivering goods or services.

You do not have to be a victim. The single most important factor in preventing crime is awareness — of your surroundings, your abilities and your options. You can reduce the risks that may make you a victim by practicing personal awareness; by understanding the mechanical operation of crimes and the motivation of the offenders who commit them; and by planning proactive deterrence techniques. Be bright and learn to recognize the relationship between cause and effect: that a particular series of events will unfold in a predictable way.

1.4 OFFENDERS

Offenders commit crimes. Offenders display offensive or criminal behavior. Offensive behavior often leads to a criminal act. Offenders are selfish and they do not care about your personal feelings, your property or your safety. Offenders generally target people who appear unaware, unprepared, untrained, distracted or vulnerable. Offenders use intimidation as their primary weapon until it is no longer effective — and then they use physical force.

Offenders rely upon the fact that most people are simply not prepared and have not been trained to cope successfully with a violent act — physically, mentally or emotionally. Most people do not see the need for — and have not developed — a personal response plan for coping with a violent act. Most people have not learned that applying effective personal safety techniques works in the business environment as well as outside the workplace.

Most people believe that "it always happens to someone else." Remember this! No one is exempt. It can happen to you. It happens to people just like you every day. Combating any type of crime means employing every available measure against it. If you learn how to become a "hard target", the offender will most likely leave you alone.

1.5 EMPLOYERS

Employers hire people to work within their businesses. The employer's primary motives for being in business are to provide a product or service, and to generate profit. An employer is not in business to provide an opportunity for people to be hurt while performing the tasks they were hired to do. Neither do employers hire people to become trained to handle circumstances that probably will not affect the business.

The problem is that employers frequently do not acknowledge the possibility that a violent act might affect an employee — and the business. The reality is that one of every six violent acts occurs within a business. The bright employer recognizes that it is not a matter of if — but a matter of when — an employee or the business will become a victim. The bright employer plans for that possibility.

1.6 WORKPLACE VIOLENCE STATISTICS

The Centers for Disease Control have officially declared workplace homicide to be a disease epidemic. Like a virulent disease, murderous violence — once considered to be a "public" crime — is rapidly becoming a "private" workplace crime. Although we most often equate workplace violence with murder, other types of violence at work are even more pervasive. There are many different categories of workplace violence, but the most serious one is homicide.

Workplace homicide is the fastest growing type of murder in the United States. Recent statistics indicate that an average of three people are murdered on the job every working day. These same statistics also indicate that:

- The most common violent workplace crimes are simple assaults, aggravated assaults, sexual assaults and robbery.
- Of offenders who make threats in the workplace, most threats come from current or former employees, followed by customers and strangers.
- Men are more likely than women to experience a violent crime.
- Men are more likely to be attacked by a stranger and women are more likely to be attacked by someone they know.
- More than 15% of violent victimizations that occurred while the victim was working resulted in physical injuries. 10% of those injuries required medical care.
- Nationally, workplace homicide is the leading cause of death at work for women, and the third leading cause of death at work for men. In some individual states, workplace homicide is the leading cause of death at work for both men and women.

- One in every six violent crimes occurs in the workplace.
- Six in every ten incidents of workplace violence occur in private companies.
- Almost one million violent crimes occur in the workplace every year.
- Another six million people are threatened and sixteen million people are harassed annually.
- Workplace victimization costs more than 500,000 employees 1,751,100 days of work each year, an average of 3.5 days per crime. These victimizations result in almost 160,000 injuries per year. Missing work because of the crime results in more than $55 million in lost wages annually, not including days covered by sick and annual leave.
- Among only those persons injured by a crime victimization at work, an estimated 876,800 workdays are lost annually, costing employees more than $16 million in wages, not including days covered by sick and annual leave.
- Victims at work are just as likely to face armed offenders as those victimized while not working.
- An armed offender faces 30% of all violence victims.
- Most work-related homicide deaths occur in retail industries, followed by service industries and public administration.
- While government employees make up nearly 20% of the total U.S. work force, 30% of the victims of workplace violence are federal, state or local government employees. Several factors may be responsible for this apparent over-representation, including a potentially high risk of victimization for particular government occupations such as public safety personnel.
- More than 50% of acts of workplace violence are not reported to the employer, or by the employer to a law enforcement agency, primarily because the victim or the company believes the incident is a private or minor matter.

1.7 THE CAUSES OF WORKPLACE VIOLENCE

Surveys conducted by research companies, law enforcement agencies, private foundations and other groups suggest that economic, cultural, legal and political changes are primarily responsible for the marked increase in workplace violence activity.

In other words, the increase in violent acts committed in the workplace are simply a reflection of the prevailing attitudes evident in today's society. It is unreasonable for an individual employer to address the primary causes for violent behavior in society. The employer does, however, have a legal obligation and an obvious self-interest in protecting employees and other persons on the company's premises.

The only things certain about workplace violence is that the threat is real and the problem is now. The variables of time, place, circumstance and people ultimately determine the cause and effect of a crime, and we will likely debate the causes for the upsurge in workplace violence for years.

None of the established reasons for the increase in violent activity is the root cause of the problem, but all of these reasons contribute to the problem:

- The Economic Connection.
- The Society Connection.
- The Work Environment Connection.
- The Legal System Connection.
- The Media Connection.
- The Substance Abuse Connection.
- The Domestic Connection.
- The Criminal Justice System Connection.
- The Gun Control Connection.

1.8 THE ECONOMIC CONNECTION

The nation's economy is unstable, creating anxiety and fear in people who are afraid of wage and hour cutbacks, or termination. If the fear becomes desperation, employees tend to verbalize their fears more — especially toward their supervisors — and often that expression reaches a point where the supervisor feels threatened.

Anxiety about losing a job or being demoted can negatively affect our professional, financial and family lives, and threaten our self-esteem. An unpredictable economy creates a more competitive, challenging and volatile atmosphere.

1.9 THE SOCIETY CONNECTION

The workplace is a microcosm of society as a whole and the number of acts of violence is increasing in all segments of society. Logically, then, an increase in violent acts in society in general means an increase in violent acts in the workplace, in particular.

1.10 THE WORK ENVIRONMENT CONNECTION

Today's work environment has changed dramatically. Job security for many people is only a memory and job insecurity for those same people has become the reality. Insecurity in one area of a person's life naturally leads to insecurities in other areas — mental, physical, emotional, spiritual and financial. A company can no longer be certain of its future nor can it reassure its employees of job security or future employment.

1.11 THE LEGAL SYSTEM CONNECTION

America's workers are filing internal complaints and suing each another for acts of real or perceived harassment. In our personal lives, we generally live in a very permissive and private society. At work, however, our behavior is regulated and scrutinized carefully to insure that we do not create situations that may threaten other persons, or that may bring a legal action against the company.

Harassment complaints and lawsuits, particularly those involving sexual harassment, often create a tense environment within the workplace. Recent statistics indicate that nearly 16,000 sexual harassment lawsuits are filed annually — and this figure probably represents only 10% of the actual claims, if we count the claims that are "handled" internally.

1.12 THE MEDIA CONNECTION

The media — audio, video and print — has brought a sense of immediacy to life. Goals must be achieved now — in this moment — and without much work. We often feel compelled to take shortcuts and to make quick decisions — which may result in permanent solutions for what are, in fact, temporary problems. Movies and television programs often glorify violence and picture it as just another tool to get quick results. They show us how to perform violent acts, and make it appear as though using violence to achieve an end is an acceptable practice. Casually and remorselessly murdering people who stand in our way is depicted as being preferable to reasoning with them.

1.13 THE SUBSTANCE ABUSE CONNECTION

Recent statistics indicate that more than 25% of the nation's population will over-consume alcoholic beverages and prescribed medications, and/or consume illegal drugs in any year — at home and at work. A person who consumes mind and mood-altering substances is unstable at best and dangerous at the worst. A person who is under the influence of such a substance often increases his/her sense of power, self-importance and physical strength — while morals, guilt, a sense of responsibility and judgment are simultaneously impaired.

People who abuse controlled or illegal substances are at considerable risk of becoming either the victim of a violent act or the offender. Substance abusers generally practice unsafe living habits — on and off the job.

1.14 THE DOMESTIC CONNECTION

Generally, you only have to look in one of two places if you want to find someone: at home or at work. These are the places where we spend most of our lives and where we have developed most of our relationships. People who are frustrated, angry or violent toward one another in one of these two places often duplicate their behavior at the other place. That means that problems that begin in one place are often played out in the other.

The workplace is the logical culmination point for violent acts involving a threat delivered by telephone or in person; a physical confrontation (battery) and an accompanying sexual assault; or stalking and a related assault or kidnapping attempt. The Bureau of National Affairs recently estimated that the cost of domestic violence to the nation's companies approaches $5 billion annually.

1.15 THE CRIMINAL JUSTICE SYSTEM CONNECTION

All participants in a violent event generally have contact with one or more components of the criminal justice system: law enforcement, the courts or corrections. Many people believe that the criminal justice system does not meet the needs of the community. These same people generally believe that if an offender commits a crime, law enforcement agencies are not effective; the courts are too lenient; and probation officers are not properly monitoring the offenders.

1.16 THE GUN CONTROL CONNECTION

The gun is the weapon of choice in more than 80% of workplace homicides, but it is not the only weapon available to an offender. Firearms are easy to buy or steal; ammunition is readily available; and regulations can be lax or nonexistent.

Acquiring a firearm by illegal means is often easier for an offender than it is for a citizen to acquire one by legal means. Restrictions on the purchase of firearms may prevent some crimes, but a person intent on committing a violent act has many alternatives for doing so.

Many cities have enacted strict laws governing the carrying and use of firearms by citizens. Many of the same cities still have some of the country's highest violent and other crime rates, years after enacting the laws.

1.17 SUMMARY

Employers, managers and employees can reduce the possibility of workplace violence by treating coworkers and customers with common courtesy, and by observing a policy of "zero tolerance" for violence. Violence in any form, from harsh and provocative language to physical contact, should never be tolerated in the workplace. If rude and contentious behavior at work becomes universally unacceptable, workplace violence will no longer be the problem it is today.

[This page intentionally left blank]

2.0 VICTIMS AND LOSSES

2.1 OVERVIEW

Everyone involved in a violent act may potentially become a victim. A victim's life is often changed forever — in ways that people who have not been victims frequently overlook. Understand that any crime can become violent and that you can feel violated — physically, mentally or emotionally. Differentiating between crimes that do and do not initially involve a violent act is useful for statistical purposes only.

Most people are only aware of the physical damage that victims suffer. Unless you have been a victim yourself, you may not understand how becoming a victim affects your emotions, your self-esteem and your self-worth. You may not recognize the hurt that results from the changes in the way your family and friends feel about you. You may not understand the negative affect that violence may have on your morals, ethics, productivity and self-reliance.

The fact that most crimes of violence are preventable is of little consolation to victims. If becoming a victim concerns you, think about the areas of possible risk in your lifestyle. Then devise a simple plan for handling or avoiding those risks — and possibly changing your lifestyle. This is not being paranoid — it is simply becoming aware of potential dangers and being prepared for them. Develop problem-solving scenarios and share them with the people with whom you live and work. Use scenarios derived from actual events and discover that the victims seldom had a plan for protecting themselves — and that following a plan would have probably prevented the event. You can benefit from their mistakes and develop a plan to protect yourself.

2.2 VIOLENCE ISSUES — PLANNING

Most people define a violent act as a simple assault committed by a stranger, with or without provocation. Violent acts actually occur in one of three ways: by design, by mistake and by misunderstanding. Acts of violence happen:

- By design: as the offender's primary goal (e.g., a terrorist act, gang initiation rite or during the permanent resolution of a relationship); and
- By mistake or misunderstanding: as an unanticipated result of another action (e.g., during a burglary, robbery or an argument).

A person without a plan and a ship without a rudder have much in common — each is at the mercy of events beyond their control. Risks avoided are risks prevented — plan to avoid potentially hazardous situations.

Consider the following variables when evaluating your areas of risk:

- **WHO**: Who is likely to become involved in an event (e.g., self, offender, family member, coworker, stranger, police)?
- **WHAT:** What is the offender's likely goal, or motive (e.g., take your money, your car, commit a sexual assault)?
- **WHERE:** Where is the offender likely to commit the act (e.g., parking lot, grocery store, home)?
- **WHEN:** When is the offender likely to commit the act (e.g., driving to work, lunch break, nighttime)?
- **WHY:** Why does the offender want to commit this act (e.g., revenge, get money, get transportation)? and
- **HOW:** How will the offender likely commit this act (e.g., firearm, car, threat only)?

Every person — including you — should develop some simple techniques for coping with a potentially violent incident. Victims who have survived violent acts generally report that they survived by making logical decisions, rather than decisions based upon the emotional aspects of the event. A logical decision is one based upon reasoning, and an emotional decision is one based upon feeling. The same survivors generally report having developed a defensive plan of action — a strategy for self-preservation. For our purposes:

- A strategy is an approach or long-range plan (policy) carefully designed to resolve a real or anticipated need.
- A tactic is an effective method or technique (procedure) for implementing the strategy.

There is no "right" or "wrong" behavior in responding to a violent act. There is only "appropriate" or "inappropriate" behavior. Appropriate behavior works. Inappropriate behavior does not work — or makes the situation more hazardous. Think about it. Developing a plan, even if you never use it, will help you to avoid surprise and to decide — in advance — to:

- Fight: to defend yourself.
- Flee: to escape; or, if you cannot either fight or flee.
- Freeze: to become immobilized, and neither fight nor flee, and to be at the mercy of the offender.

We are all afraid of something — or someone. It is natural to fear an event that you have never encountered before, and to fear a violent person that you have encountered before.

Learning to acknowledge and to control your fear, and to channel the energy you spend being afraid, is part of the survival process. A logical response helps you to control your fear and your other emotions until the event is over. The types of incidents — which frequently overlap — that most often result in a person becoming the victim of a violent act are a(n):

- Simple assault
- Sexual assault
- Burglary
- Robbery
- Extortion
- Relationship
- Terrorist act
- Random act

2.3 SIMPLE ASSAULT

The act of making physical contact with you without your consent and often without warning. The offender's primary goal is to express fear, frustration, anger or to dominate you.

2.4 SEXUAL ASSAULT

The act of forcing you to participate in a sex act without your knowledge, permission and/or consent. The offender's primary goal is to dominate, humiliate and degrade you.

2.5 BURGLARY

The act of entering a building intending to take property or to commit some other crime. The offender's primary goal to obtain your property — primarily your money and other valuable personal effects.

2.6 ROBBERY

The act of taking property from your person or your presence, against your will, by using force or fear. The offender's primary goal is to obtain property — primarily your money and other valuable personal effects.

2.7 EXTORTION

The act of obtaining your property, with your consent, by using force or fear. The offender's primary goal is to exchange something you value — a person, thing or some information — primarily for money.

2.8 RELATIONSHIP

The act of terminating a relationship by using force or fear, or of using violence as a means of revenge. The offender's primary goal is to use threats or acts of intimidation, vandalism, sabotage, extortion and terrorism to cause the other person in the failed relationship — you — to comply with his/her demands.

2.9 TERRORIST ACT

The act of gaining recognition for a personal or political cause or action, by means of force or fear. The offender's primary goal is to use threats or acts of intimidation, vandalism, sabotage, extortion and terrorism to focus upon his/her specific demands.

2.10 RANDOM ACT

The act of committing a violent act, seemingly for no apparent reason. The offender's primary goal is to commit a shockingly visible and witnessed violent act that enhances his/her image, or to gain personal recognition for a personal or political cause or action. This type of violent act is committed for the "thrill", "rush", or "experience", and it is often committed as a gang initiation rite.

Most threats of physical violence made during an event result in actual injury to a victim. Violence usually occurs when:

- The offender confronts the intended victim.
- The offender has not thoroughly planned the act.
- The offender wants no witnesses.
- The victim has not planned for his/her response to the act.
- An event occurs that alters either the offender's or victim's planned actions or response.

If you have never thought about how you would react in such an emergency, it is likely that you will make inappropriate decisions. Responding effectively to emergency situations requires that you plan carefully and receive training that involves critical decision-making. There are two critical decisions that every person should make before a potentially violent event occurs. These decisions will ultimately affect your strategy, tactics and options for coping with violent acts:

- Am I willing to fight, to be injured or to sacrifice my life for another person — and if so, under what conditions?
- Am I capable of injuring or killing a person to defend my life, or the life of another person — and if so, under what conditions?

The potential opportunity for most violent incidents to develop may be reduced, and your decisions about these two questions may be simpler, if you remember these simple guidelines:

- Understand that you are unable to help anyone else unless you first help yourself. Think of your own safety, well-being and actions before considering any another person's.
- Consciously practice prevention in your daily activities so that you establish and maintain safe living habits. Avoid potentially hazardous locations, events and people.
- Participate in a self-defense training program to develop your basic physical skills, mental balance and self-confidence. Your display of a self-confident attitude is often sufficient to repel an offender.
- Keep your plans and actions simple to avoid mistakes and confusion. Plan as if everything will go wrong; that help will be unavailable; and that your survival depends solely upon your own efforts.
- Discuss your plan with people you live and work with if your response may affect their plans. Survival often includes both individual effort and teamwork.

2.11 VIOLENCE ISSUES — ACTION

Planning appropriately and practicing prevention behavior is often sufficient to cause an offender to "target" someone else. Plan to act appropriately during hazardous situations if avoidance behavior fails, and to apply the same methodology as you used in the planning section:

- **WHO**: Who is likely to help me (e.g., self, family member, coworker, stranger, police)?
- **WHAT**: What tools do I have available to help me (e.g., car horn, flashlight, keys, spray repellent)?
- **WHERE**: Where can I go for help (e.g., police or fire station, any store, home)?
- **WHEN**: When is the best time to act (e.g., before the assault begins, during the assault, now)?
- **WHY**: Why do I need to act (e.g., self-preservation, protect family members, distract others)?
- **HOW**: How can I best survive this event (e.g., cooperate, use non-lethal force, use lethal force)

The potential opportunity for you to survive most violent incidents may be enhanced if, in addition to the prevention guidelines, you remember these simple action guidelines:

- Report suspicious and unusual events, conditions or people to the police as soon as possible. No one can help you unless they know that you are in trouble.

- Treat every threat as real and every offender as dangerous — and capable of carrying out the threatened action. Underestimate your abilities and overestimate your adversary's.
- Assume total responsibility for the event and do not attempt to involve other persons if possible. Your risk increases when you involve or rely upon other people in the event.
- Train your mind first. Your body will follow the directions and commands from your mind. If your body receives no directions or commands from your mind, it will freeze and you will be unable either to fight or flee.
- Understand that defenders never win. The only hope you may have is to attack your attacker, and you must change your focus from defense to offense. Sometimes we win; sometimes we lose; and other times we must be content simply to cut our losses.

2.12 THE PSYCHOLOGICAL IMPACT

After a violent incident occurs, people often experience many emotional and behavioral changes. These changes can result in intense fear; a sense of vulnerability; and confusion and disorientation concerning time, place and persons. These changes are normal — expect them. These are normal responses that normal people have when exposed to an abnormal event.

Victims experience a predictable range of emotions after the event — denial, rage, anger, frustration, grief and finally — acceptance. Victims have to experience each of these stages so that the healing process can take place.

Victims often report experiencing nightmares and having trouble sleeping. Victims often experience an overwhelming sense of guilt, feeling that they should have done something that would have prevented the event. A victim's distress often manifests itself in physiological problems — frequent illnesses and "accidents".

After a violent event, victims often wonder why they interact differently with their coworkers or why they have difficulty at home with their spouses and their children. They may feel depressed and out of control, and they may ask themselves "What's wrong with me?" and "Why am I feeling this way?"

2.13 THE PERSONAL VICTIM

A person who becomes a victim experiences a personal and individual reaction to the event. Victims personally:

- Feel a sense of loss.
- Experience an unnamed fear.
- Are not productive.
- Have to pay medical costs.

2.14 SENSE OF LOSS

Victims generally experience an overwhelming sense of undefinable loss, often losing a sense of self and, in the most severe cases, losing sight of all reasons for living. In addition to feeling violated, victims experience a loss of control — generally manifested as a lack of self-control.

2.15 AN UNNAMED FEAR

A victim's greatest fear becomes: what is to prevent this from happening again? Added to this is the uncertainty of what may have happened to the offender since the event. The victim often experiences an unreasonable increase in fear — taking fear to the next level and becoming paranoid.

2.16 LOSS OF PRODUCTIVITY

Victims become dysfunctional, often becoming physically and emotionally unable to even the most basic daily tasks. They become preoccupied and inattentive, causing them to make mistakes and to perform at less than their peak levels.

2.17 MEDICAL COSTS

Becoming a victim is expensive. Medical costs for the initial treatment of physical injuries are often very significant. But it is the continuing cost for outpatient care and counseling for victims, their families and for other affected employees that causes the true cost of the event to escalate dramatically.

2.18 THE COMPANY VICTIM

The company is a victim of these incidents, too — perhaps even if the event did not occur on company property. In addition to addressing the personal issues that arise within employees, the company as an entity will experience:

- Higher insurance costs.
- The need for retraining.
- An increase in business losses.
- A loss of image and reputation.

2.19 HIGHER INSURANCE COSTS

If an employee is injured at work — regardless of the cause — the employee will probably file a worker's compensation claim. These claims are often filed months after the original incident and they can be very expensive.

Health insurance costs can be affected as well. Health insurance companies base the cost of a group policy upon the number of claims the company files annually. Property and casualty insurers base their rates upon the number of reported claims. Companies offering director's and officer's liability insurance base their rates upon the company's exercise of "reasonable and prudent care" in conducting its business activities. One act of workplace violence may affect all policies significantly, adversely and simultaneously.

2.20 NEED FOR RETRAINING

Any event that identifies some deficiency in design, planning, preparation or implementation of a company's policies and procedures may cause new policies and procedures to be developed. This will naturally cause a cycle of retraining at all levels within the company. If employees leave the company because of the event the company will then have to train new employees.

2.21 INCREASE IN BUSINESS LOSSES

The hidden losses generally peak in about six months and a full recovery to normal operations may take three years. These losses include:

- An increase in the company's expenditures for premises clean up, repairs, possible facility move and improved security measure.
- Lowered employee morale due to increased security measures and aggravated by distrust and resentment of other employees.
- Employees' disorientation and resulting mistakes, causing poor customer and interdepartmental service.
- A reduction in actual tasks performed by employees because of their preoccupation with the event.
- Disruption of normal operational activities as employees are re-allocated to perform investigative and security tasks.
- Salary and benefit continuation for injured or traumatized workers, and for workers who quit or retire as a result of the incident.
- Adverse publicity due to management's inability to control the flow of information to the media.
- The necessary interaction with the criminal justice system.

2.22 LOSS OF IMAGE AND REPUTATION

A company's image takes years to acquire, daily effort to maintain and one negative event to lose. People are generally much better at remembering one adverse circumstance than remembering thousands of good deeds.

Hiring good people is difficult if your company has a negative image. Good employees may not want to stay with a company that receives negative publicity. The attitude and morale of employees can be affected for years by violent events. For instance, just mentioning workplace violence to your postal carrier can elicit a very negative response.

2.23 SUMMARY

Workplace violence affects us all. The price we pay for everything we buy is driven higher by the costs of violence at work. And what about our image as a people? We are considered a violent society by much of the civilized world. How much does that cost us in international business and mutual trust? In a very real way, we are all victims of violence in the workplace.

[This page intentionally left blank]

3.0 IDENTIFYING OFFENDERS

3.1 OVERVIEW

In the world outside of the workplace, people of all kinds commit violent acts. Some offenders target specific victims — most offenders do not. Victims of violent acts are generally victims of opportunity. They are often simply in the wrong place at the wrong time — victims of time, place and circumstances.

Violent acts within or related to the workplace are much more specific, more deliberate and more personal. Many acts of workplace violence involve people that we work and/or live with every day. We know them and they know us.

No one does anything without a motive. A motive is simply a reason for behavior. The primary motives for offenders, both business and non-business, to commit any violent act, are:

- To dominate, degrade and humiliate the victim — to compensate for feeling inferior, or "less than".
- For revenge or in retaliation for a real or imagined injury.
- For financial gain.

3.2 WHO ARE THE OFFENDERS?

According to the FBI and the Census of Fatal Occupational Injuries, there are several categories of violent offenders, including a(n):

- Disgruntled co-worker or former employee.
- Person committing robbery and other types of related crimes.
- Irate customer or vendor.
- Person who is romantically obsessed or fixated upon a particular target.
- Person who commits random acts of violence for the thrill.

These same statistics indicate that workplace homicides are committed most frequently by people in these categories:

- Homicides by customers — 24%.
- Homicides by strangers — 44%.
- Homicides by co-workers — 20%.
- Homicides by supervisors — 7%.
- Homicides by former employees — 3%.
- Homicides by family members and others — 2%.

Responses to interviews conducted after the crime was committed, and observations from event participants, indicate that the reasons (motivations) offenders assaulted their victims were as a result of:

- Events occurring during a robbery.
- Non-specific and generally non-related personal problems.
- Dissatisfaction with service.
- An interpersonal conflict related to employment.
- A specific reaction to a disciplinary action not resulting in termination.
- An opportunity or perceived necessity discovered during the commission of another crime.
- A layoff or termination.
- A victim-specific hate crime.

3.3 IDENTIFYING POTENTIAL OFFENDERS

Identifying potential offenders of any kind of workplace violence — from verbal threats to physical assaults — is done by evaluating the person's past and present behavior and other stress factors. Certain types of behavior and stress factors are frequently indicators of potential violence. When these indicators are combined, they form a "profile" of the potentially violent person.

While profiles are effective, they are not always accurate. A person fitting a profile may not commit a violent act, and someone who does not fit a profile may become violent — but those are the exceptions. Profiles generally are very effective. In fact, about 95% of people who "fit" violent offender profiles are, in fact, dangerous. A profile simply helps to give us a population that we need to look at more closely.

An important part of a profile is a person's past behavior and stress factors — the person's "case history". Some of this information may be found in personnel files. More information may be found in medical, criminal or military records. Much of the information may be confidential or protected by law. But sometimes potential offenders volunteer information about their history to co-workers, supervisors or others. If this information is reported and handled sensibly, it can play a key role in avoiding a tragic incident.

3.4 HISTORY OF SIMILAR BEHAVIOR

People raised in this type of environment generally have an obsessive need for order and structure. If these needs are frustrated, if the work environment becomes chaotic, or if either work or home situations become unstable or uncertain, these persons have significant difficulty coping with everyday events.

There is also a direct correlation between those offenders who engage in domestic violence and those who engage in workplace violence. Frequently these types of offenders have also been victimized through mental, emotional, physical or sexual child abuse.

3.5 HISTORY OF MENTAL ILLNESS

Most workplace violence offenders have experienced some form of mental illness — minor, moderate, or severe. These offenders are generally angry with something or someone; they are resistant to change and the accompanying loss of personal control; and they may experience delusions and hallucinations.

In the worst case, these offenders begin to see the people around them as things — things that their rage makes them want to humiliate or destroy — things that they want to devastate. It is this destructive impulse that often drives these offenders to commit violent offenses.

3.6 HISTORY OF PHYSICAL PROBLEMS

Sometimes there are physical or organic reasons why people become violent. If a person develops a brain tumor in the vicinity of the brain's center of emotion, for example, the resulting physical impairment may create violent changes in the person's moods and reasoning capabilities. The accompanying emotional result may be displayed as intense outbursts of rage that are not motivated by any external influence — and that seemingly come from nowhere.

Physical problems may result from some injury to the central nervous system that happened prenatally, later in life, or through illness or accidents. These injuries can also affect the individual's impulse control capability. Physical imperfections or deformities may also cause a person to develop an angry or resentful disposition.

3.7 EVALUATING VIOLENCE POTENTIAL

It is not sufficient to consider only the person when we try to evaluate the potential for violence in the workplace. We must evaluate the person in connection with the environment and the "triggering event".

3.7.1 Environment

Some workplace environments are more conducive to violence than others — the so-called "toxic workplace" — where the environment has a history of potentially violent situations. Examples of these environments are worksites that experience:

- Chronic labor and management disputes.
- Frequent grievances filed by employees.
- High levels of stress due to the nature of the work performed.
- Extraordinary numbers of injury claims — especially psychological injuries.

- Understaffing or excessive demands for overtime placed on staff.

3.7.2 Triggering Events

Triggering — or "catalyst" — events are the combination of factors that provoke the offender to act. These events vary depending on whether the offender is a co-worker, a customer or a stranger. If the offender is already mentally or emotionally unstable, the introduction of a new element — or incident — may trigger behavior that is inappropriate to the situation, such as if the offender:

- Becomes embarrassed about making a mistake and begins shouting profanities at co-workers.
- Becomes angry with a computer because it is slow and destroys it.
- Becomes frustrated with a company vehicle that will not start and shoots it.

3.8 PREDATOR PROFILE

We often call violent offenders human "predators" because they prey upon their victims. They hunt — or stalk — their victims and then "feed" on the victims' pain and suffering. Human predators generally enjoy the hunting "game" and often "toy" with victims before causing any serious injuries, just like their counterparts in the animal kingdom. Both human and animal predators generally become more ruthless and more deliberate with each successful "hunt". With each new success, each type of predator becomes emboldened to take new risks.

Offenders who commit violent acts in the workplace frequently broadcast signals — behaviors — that they will be more prone to violence in the workplace than other people. These behaviors, also known as "crisis signals", are often displayed both at work and at home simultaneously, in a pattern of escalation that suggests an emerging and evolving violent personality.

Displaying one or more of these behaviors does not mean that the person will be violent, but these behaviors are frequently displayed by persons "at risk" to committing acts of violence. The more behaviors that a person displays equates to the more likely the person will commit a violent act. These persons may be co-workers, family members, customers or vendors. A person is most at risk when he/she:

- Has a history of violent behavior, either on or off the worksite. This is the most significant predictor of future violence. Examples are incidents involving spousal abuse, child abuse or cruelty to animals.
- Pays decreasing attention to personal hygiene or appearance. Examples are the failure to wear clean clothes to work, failure to bathe regularly or to maintain one's facial appearance.

- Displays a willingness to compromise personal values, morals, ethics and personal safety — remaining just within acceptable limits. An example is taking unwarranted safety risks and encouraging or daring others to take similar risks.
- Is obsessively involved with the job, often with no apparent outside interests, few friends and no close family — and/or relationships that are constantly in turmoil. The danger is that work becomes the person's sole source of identity.
- Displays contempt for the company and its leaders, and repeatedly and willfully violates company policy and procedures. Examples are making derogatory and insulting remarks in private — and sarcastic remarks in public — about supervisors, and daring a supervisor to enforce a policy.
- Has a high rate of tardiness, absenteeism and often fails to complete assigned work projects on time. Examples are using excessive sick leave or previously-unscheduled vacation time; failing to appear for meetings at the scheduled time; unapproved or unusual disappearances from the workplace; and offering inappropriate or improbable excuses for absences.
- Negatively impacts the supervisor's time and energy. Examples are repeatedly requiring special coaching or motivation to perform ordinary tasks; repeatedly requiring special counseling to address personal problems not relating to employment; and causing repeatedly events that must be documented for potential disciplinary action.
- Displays noticeably inconsistent work patterns and significant concentration problems. Examples are experiencing alternating periods of low and high productivity; inability to give or follow instructions; chronic forgetfulness; and on-the-job injuries indicating distraction.
- Is a loner with little or no involvement with co-workers, or he/she develops an uninvited romantic interest in a co-worker. An example is when the romantic interest becomes an obsession and the offender's words or actions threaten the co-worker, who may report the attention as harassment.
- Routinely consumes mind- and mood-altering substances — alcohol and drugs — as a means to enhance his/her self-esteem and self-worth, or to reduce guilt and emotional pain. Mind- and mood-altering substances interfere with reasoning ability, social inhibitions and the ability to distinguish between right and wrong. Examples are drinking alcoholic beverages during the meal break or arriving at the worksite intoxicated.
- Experiences depression and desperation over current or recent domestic, financial or personal problems — and often blaming another person or event while failing to accept personal responsibility for the results. An example is the unwillingness to accept that a failed relationship has ended.

- Acts paranoid, feels pursued, displays extreme mood swings and over-reacts to ordinary events. An example is when an employee states that someone is "out to get" him/her, unsubstantiated by facts.
- Does not accept constructive criticism or personal responsibility for results, blames other persons or circumstances for mistakes or poor performance, and does not forget what he/she interprets as an "injustice". An example is insisting upon being "right" while making you "wrong".
- Shows an unusual interest in — or obsession with — weapons. This signal is particularly significant if it starts spontaneously or is out of character for the employee. Examples are developing an evolving interest in paramilitary training or groups; accumulating weapons collections, including semi-automatic weapons; and compulsively reading and collecting gun, explosives or paramilitary magazines.
- Talks excessively about weapons at work, carries a concealed weapon or displays a weapon to "test" a person's reactions. An example is observing a weapon lying on the top of an employee's desk.
- Intimidates others or instills fear in co-workers or supervisors — either physically or verbally. Examples are making harassing telephone calls, bullying or stalking.
- Makes verbal threats, either directly or indirectly. An example is predicting that "bad things" are going to happen "soon" to a co-worker or a supervisor.
- Makes spontaneous statements concerning bizarre religious or political beliefs, or makes "doomsday" predictions. These statements are particularly significant if the employee is adamant that these are the only "true" beliefs. An example is making a statement that a "righteous God will destroy all nonbelievers."
- Makes it known that he/she is anticipating death and makes a display of putting his/her personal affairs in order. Examples are closing bank accounts, writing a will and arranging for prepaid burial services.
- Is fascinated with other recent incidents of workplace violence, frequently accompanied by voicing verbal approval of that offender's actions. An example is observing a newspaper clipping of the incident displayed prominently in the employee's work area.
- Becomes attracted to, fixated upon or talks frequently about explicitly violent or sexually violent materials. An example is the exclusive reading of violent fantasy comic books or maintaining a pornographic, sado-masochistic magazine collection.
- Becomes inappropriately distressed during workplace events that cause stress. Examples of these events are accelerated production schedules and forced overtime assignments; notices of pending layoffs or terminations; and work reductions or labor disputes.

- Works at a location where employees are exposed to the threat of violence from non-employees. An example is having to drive through a neighborhood controlled by a gang in order to travel to and from work, or working next to a business that is a target for activists' demonstrations.
- Fails to take appropriate and consistent disciplinary action against employees that he/she supervises, when they threaten to or commit acts of violence against other employees. An example is failing to properly investigate a sexual or other harassment claim.

3.9 BATTERER PROFILE

Domestic violence is not just a "domestic" issue. Because the domestic violence "lifestyle" is frequently duplicated in the workplace, a domestic batterer is likely to exhibit similarly-violent behavior at work. More that 90 percent of all domestic violence offenders are male, as are the majority of workplace violence offenders. The violent behavior of a spousal and/or child abuser is very similar to the behavior of human predators. The domestic batterer often displays many of the same behaviors as the predator and creates an environment that is increasingly controlling, manipulative and fearful.

Understanding the correlation between the domestic batterer and the workplace violence offender is important in preparing prevention and crisis management strategies. Put yourself in the abused victim's role to gain a better perspective of this unique — and very dangerous — personality. In addition to exhibiting many of the predator behaviors, the batterer is generally:

- Blameless (e.g., the abuser refuses to take personal responsibility for failed relationships, careers or the lack of attainment of life's goals — and blames you instead).
- Selfish (e.g., the abuser uses the words "you" or "they" to describe failure instead of the word "I", and uses the word "I" to describe success and does not include the words "you" or "they".
- Hypersensitive (e.g., the abuser is easily and frequently upset by annoyances, inconvenience and criticism — particularly caused by or coming from you).
- Argumentative (e.g., the abuser insists on being "right" while telling you that you are "wrong").
- Unpredictable (e.g., the abuser displays sudden and marked mood swings and unpredictable behavior).
- Has a history of abuse (e.g., the abuser has a record of battering and other abusive behavior, and has probably also been a childhood victim of similar abuse).
- Jealous (e.g., the abuser resents other people — particularly you — spending time with their relationships — family, friends and co-workers).

- Controlling (e.g., the abuser makes you account for where you go, how you spend your time and how you spend your money).
- Manipulative (e.g., the abuser removes you from the decision-making process regarding important issues in your life, often explaining to you that you are "too stupid" to make your own decisions).
- Isolated (e.g., the abuser deliberately and systematically cuts you off from all supportive resources and influences, including your family, friends and co-workers).
- Physically violent to objects (e.g., the abuser often breaks your — and only your — possessions);
- Cruel (e.g., the abuser is insensitive to or intentionally inflicts pain and suffering upon you, children and animals).
- Verbally abusive (e.g., the abuser says cruel and hurtful things to degrade and humiliate you — particularly in front of other people).
- Threatening (e.g., the abuser tells you he/she will hurt or kill you — and often does it).
- Physically forceful (e.g., the abuser holds you too tight, kisses you too hard and insists on having sex whether you want to or not).
- Physically abusive (e.g., the abuser pushes, shoves, slaps, kicks or hits you).

Again, exhibiting some of these behaviors on an infrequent basis does not necessarily mean that a person is a batterer. The overall pattern of behavior of a potential offender and his/her environment must be assessed to evaluate his or her "violence potential" effectively.

3.10 SUMMARY

Abrupt changes in behavior are often just as telling as any pattern. A rude, harsh and loud-mouthed person who suddenly becomes quiet, devious and sinister may be more dangerous than a person who is habitually boisterous and offensive.

A courteous, tidy and punctual person who becomes sullen, unkempt and tardy may become a significant risk. It is important that someone, and not necessarily you, talk to these people or to their supervisors. This pro-active approach may get them some help in order to avoid a "festering" problem, which might erupt into violence later.

4.0 WHAT EMPLOYEES CAN DO

4.1 OVERVIEW

Violence is brought to the workplace by two types of offenders — those who have a connection to the business and those offenders who do not. Employees and ex-employees; customers and ex-customers; managers and ex-managers; current and former romantic partners — these people have a connection to the business.

Robbers, burglars, rapists, gang members, terrorists and other criminals generally have no connection to the business. Understanding the motives of these two groups can help employees recognize the danger signs that may precede violence. The possible combinations of employee and non-employee are:

- Employee vs. employee (includes your supervisor).
- Employee vs. non-employee (includes customers, vendors and strangers).
- Non-employee vs. employee (includes your family members).
- Non-employee vs. non-employee (includes all other persons).

4.2 CONNECTION TO BUSINESS

Offenders with a connection to a company present an interior threat. Sometimes, interior threats are easier to detect than exterior threats. If you learn what to look for, both employees and supervisors can often recognize the danger signs of impending violence in co-workers and in their other relationships.

The motives of offenders with a connection to a business generally fall into the categories of revenge or jealousy. A disciplinary action, being fired, a poor performance review, the lack of promotional capability, the loss of a requested pay raise or having health problems — combined with other types of stress — can cause some people to feel justified in committing vengeful acts. The offender generally does not care who is hurt in the selfish pursuit of revenge.

Jealousy is often a powerful motive for committing a violent act during a marital separation, a divorce, an unrecognized occupational accomplishment or the breakup of a romantic relationship. Domestic issues culminating in acts of violence at the workplace continue to increase — for one very simple reason. A domestic partner may break off a relationship with an abuser and change living locations, but changing jobs is harder — so the abuser knows where to find the victim during work hours.

Customers who feel that they have been offended, treated unfairly or cheated, can become so enraged that they resort to acts of revenge — including verbal abuse, a harassment campaign or a physical assault.

4.3 NO CONNECTION TO BUSINESS

Offenders with no connection to a company present an exterior threat. Alert and observant employees and the use of an effective Security Program can prevent most of their criminal acts.

Offenders having no connection to a company, like robbers, extortionists, muggers, rapists and arsonists are most often driven by either of two motives — financial gain or victim dominance. They want money or the feeling of power that accompanies degrading or destroying the victim.

4.4 RECOGNIZE DANGER SIGNS

Offenders generally exhibit specific danger signs that an employee can readily observe and report, including:

- Using abusive, rude, threatening or intimidating language.
- Displaying inappropriate emotional and physical reactions to events
- Obviously attempting to control their emotions.
- Adopting a suspicious appearance, particularly unusual or conspicuous apparel, grooming or behavior.
- Loitering in areas that are accessible to the public.
- Appearing in areas that are not accessible to the public.
- Making threats of any kind to other persons, including customers, employees, supervisors, managers, or against the company itself.

4.5 REPORT DANGER SIGNS

All employees, including you, are responsible for reporting potentially hazardous circumstances to their supervisors. Often employees fail to report these events because they fear embarrassment, retaliation or disciplinary action. A company has the obligation to act upon reports received and at least to determine their credibility. Reporting potentially hazardous circumstances allows individuals to plan for their safety and for the company to implement security measures to protect the reporting employees and others.

It is critical that employees report all potentially dangerous incidents that occur in the workplace, and also those events that occur in their personal lives that they could bring into the company. Every threat should be treated as if it is real until it can be proven groundless.

4.6 PERSONAL PROBLEMS

Employees can start practicing safe living and working habits by recognizing, and reporting where appropriate, the dangerous areas within their own lifestyles. Employees should advise supervisors (and appropriate co-workers) of potentially dangerous issues in their lives outside work that could potentially bring acts of violence into the workplace.

4.7 CO-WORKERS

Conflicts may begin at the worksite because of minor disagreements or misunderstandings. Supervisors and co-workers are often able to resolve these conflicts quickly and fairly — but only if they are made aware of the problem. Several employees may be experiencing similar conflicts with the same co-worker. When supervisors see an emerging pattern of conflict involving a particular employee, they should talk to that employee and resolve those issues creating the conflict, or refer that employee for more help. It is best to resolve problems early rather than letting them escalate into a major incident.

4.8 CUSTOMERS AND OTHER PERSONS

Customers and other persons on the premises, such as vendors, often present special problems. Supervisors and co-workers often do not have the same familiarity with this type of person's personality as they do with other employees. Customers and other persons are usually on company premises by invitation and, unless they break the law, their behavior is largely unregulated. However, if they create a disturbance, they can be asked to leave. If they refuse, they can be arrested for various reasons.

4.9 STRANGERS

Persons with no obvious affiliation with the company may appear on the premises to gather visual information so that they can plan future crimes. These unaffiliated persons may telephone the company to gather information that they cannot obtain visually.

General Safety Strategies

Besides recognizing and reporting danger signs, employees can adopt general safety strategies, learn to de-escalate tense situations and be prepared for possible violence. The four most important steps employees can take to protect themselves from acts of violence are observing, reporting, de-escalating and retreating, such as to:

- Observe people's behavior.
- Report anything unusual or suspicious.
- De-escalate tense situations.
- Retreat to safety if de-escalation does not work.
- Identifying your available security options is important to your safety. Before you are ever confronted with a potentially violent event.
- Enter and leave the premises in groups.
- Talk with the person who is the designated Unit Security Officer or safety coordinator for your worksite. Discuss your safety concerns and the company's emergency response plan, including your role.

- Learn what the company wants you to do in an emergency. Learn how to use alarm systems, panic buttons or other emergency systems that you may have available to you, and have emergency contact information posted by your telephone.
- Remove objects that could be used as weapons against you, from desks and other areas where you interact with customers.
- Develop an emergency signal or code word that you can use with your small work group.
- Identify an escape route and practice moving quickly from your work area in case you have to retreat.
- Acknowledge all persons present on the premises with courtesy and respect and do not give anyone a reason to feel ignored. If you cannot answer a question or resolve a situation, find someone who can answer or resolve it for you.

4.10 IMPLEMENT DEFENSIVE STRATEGY

Implementing a defensive strategy is also important to your safety. If you are confronted with a potentially violent event:

- Avoid physical contact with the offender regardless of harassment or verbal abuse, but remember that you can defend yourself if you are attacked. Self defense is a basic, personal right.
- Leave the area immediately, if it is possible.
- If escape is possible, notify your co-workers, supervisors, and others on the way to safety.
- If leaving the area is not possible, attempt to de-escalate the event. De-escalating tense situations will often prevent violent confrontations.
- If de-escalation does not work, you must be prepared for possible violence. That means retreating, if possible, or knowing what to do to minimize your chances of being hurt. If you do not know these things, you should learn them. Talk to your Security Department or your local law enforcement agency about what you can do to prepare for and respond to potentially violent incidents.
- Report verbal incidents to your supervisor immediately — and if a physical offense occurs, notify any supervisor or a law enforcement agency as quickly as possible.
- Discuss the incident only with appropriate persons.

4.11 SUMMARY

There are many ways to enhance your security at work and the best way is to be prepared. You can minimize the chances of becoming involved in a violent incident by being observant; by reporting suspicious or questionable behavior; by avoiding confrontation; and by learning how to de-escalate tense situations.

Notes

5.0 WHAT EMPLOYERS CAN DO

5.1 OVERVIEW

Businesses pay the price of workplace violence in lost time, decreased productivity, increased employee turnover, increasing medical and insurance costs, litigation, increased internal and external security and lost trade.

Developing and implementing your company's Security Program is every employee's responsibility, not just your Security Director's. One of the main objectives of a Security Program should be to establish and maintain a safe and secure environment in which employees may work and customers may conduct business.

5.2 LEGAL ISSUES

Various city and county ordinances, and state and federal laws, provide that employers have the general responsibility for providing and maintaining a safe and secure environment for employees and customers — within the realm of "foresee ability". This means that if your company has some reason to believe that a hazardous condition now exists — or is likely to exist in the foreseeable future — it must take appropriate measures to protect all persons likely to be affected.

The General Duty Clause of the Occupational Safety and Health Act (OSHA) of 1970, Sec. 5(a) states: "Each employer: (1) Shall furnish to each of his (or her) employees employment and a place of employment which are free from recognized hazards that are causing or are likely to cause death or serious physical harm to his (or her) employees..."

Employers have a general duty to provide a safe workplace. Employers should understand that OSHA not only has specific regulations that deal with equipment-oriented issues in workplaces, but they also have the regulatory authority to ensure that the employer is providing a safe workplace for its employees generally. Therefore, under the General Duty clause, OSHA inspectors are responsible for determining whether or not employers are providing a workplace that is safe from violence.

Workers' Compensation laws also state that your company must compensate your employees for injuries sustained during their employment if, the injury arises out of and in the course and scope of employment. This generally means that, if an employee is acting on your behalf in some capacity and becomes injured, your company must compensate the employee.

Your company may be liable to non-employee third persons for damages caused by your employees, who are performing their duties within the scope of their employment or otherwise acting in your company's behalf. This means that if one of your employees injures another person while working for you, that your company may be held liable for damages. If the employee is not acting within the scope of his/her employment, your company may not, perhaps, be held liable for damages. "Willful and malicious acts" are generally not considered within the scope of employment.

If an employee injures a non-customer, there likely will be several issues to be decided:

- Was the employer negligent in hiring that person in the first place?
- Was there was any foreseeable reason that the employer should have known that this employee was going to be violent?
- Was the employer negligent in supervising that employee's work?

Generally, when an employee injures another employee as a result of a violent incident, liability will be covered by the employer's workers' compensation plan. It does not matter whether the employer was negligent or whether the employer had any fault at all: the mere fact that one of the company's employees has an injury will result in workers' compensation liability in most states.

5.3 PRE-EMPLOYEE SCREENING

Employers must review their application forms carefully and determine whether the form requests appropriate and sufficient information from the applicant. Use both the written application and the interview process to inform applicants clearly that:

- Lies or omissions on the application form, discovered during the interview process or after the applicant has become employed, will be sufficient grounds for the denial of employment and/or for dismissal.
- The company maintains a zero tolerance policy for violence in the workplace — and that the applicant must sign a policy acknowledgment statement.
- Hiring and retaining individuals depends on the competency of the individual, the character of the individual and the chemistry between the individual and the company.

5.4 VALIDATE PAST EMPLOYMENT HISTORY

Employers must also look closely at the applicant's past employment, using the employee's application form and making calls to past employers and other listed references to investigate. First consult with your company's legal counsel before making these calls. Generally you can ask these types of questions and make these inquiries:

- Ask behavioral questions to verify information contained in the application and information gathered during the interview. Ask specifically behavioral-oriented questions about the applicant; and

ask for performance reviews. Employees are entitled to possess copies of performance reviews and to share them with you.
Employers should document the results of their investigations into the applicant's background, demonstrating that reasonable and prudent care was taken to insure that it was appropriate to hire the employee. Retaining the documentation of applicants who have been hired or who have not been hired is important as well, since the applicant may re-apply to your company in the future.

5.5 PSYCHOLOGICAL TESTING

Some workplace environments and job responsibilities are more stressful than others. Matching the personality "type" with the type and scope of work to be performed is important. Psychological testing, when conducted by a qualified professional, is an excellent way to pre-screen applicants for suitability. Consult with your legal counsel for advice about what employment positions should qualify for this screening technique.

5.6 SUBSTANCE ABUSE TESTING

More than 25% of this country's workers use illegal drugs or overuse prescription medications — many of them while the employee is at work. During most of the violent events committed either at work or at home, either the offender, the victim or both the offender and the victim were under the influence of a mind- or-mood altering substance. A carefully-designed drug awareness and screening program often eliminates applicants who are substance abusers before they are employed, and establishes a cause for removing current employees who are substance abusers. The substance abuse awareness program and policies regulating substance abuse should include alcoholic beverages. This type of program generally must be applied to and include all levels of employees.

5.7 ADOPTING PREVENTION STRATEGIES

Carefully-defined prevention strategies resolve most issues contributing to workplace violence before those issues become critical. The primary prevention strategies to consider include designing a crisis management plan and selecting a crisis management team; developing policies and procedures; installing appropriate security devices; constantly reviewing security procedures; documenting events; developing training programs and involving your local law enforcement agency in the development and implementation of your program.

5.8 DESIGNING A CRISIS MANAGEMENT PLAN

Conduct a risk assessment to determine what potential threats should be included in the company's crisis management plan. The assessment should be simple and should be conducted by you and your department managers. The risks should be realistic and take into account the company's history, geographical location, industry-specific trends, employee profiles and related industry statistics.

After identifying the potential risks, include a plan for responding to these events with your available resources. If you determine that your available resources are inadequate, identify the means you will use to obtain resources from outside the company.

5.9 DEVELOPING POLICIES

When you develop policies, remember that telling someone not to do something is seldom sufficient, and then telling them only once — you have to tell them what to do instead and tell them often. Employees cannot meet an employer's expectations unless you make them aware of those expectations repeatedly. Employees need to know what your expectations are, and then what the consequences are for breaching any of those expectations.

5.10 ESTABLISHING A COMPLAINT AND FOLLOW UP PROCEDURE

Establishing a complaint and follow up procedure to address employees' and customers' complaints is very important. If people know that there is a procedure for dealing with their problems, they likely will not be angered by the feeling that they have "no say" in the matter — and no means to redress their grievances. Eliminating anger and the accompanying loss of control will reduce the likelihood of violence. Establishing a complaint procedure provides a measure of fairness and protection for all parties to a relationship, and establishing a follow up procedure indicates that you will investigate the complaint.

5.11 ESTABLISHING A THREAT POLICY AND PROCEDURE

You should develop a threat policy that requires employees to report threats to any supervisor. Your company's position should be that it takes all threats seriously and that it maintains a "zero tolerance" for threats. The position that your company should take when defining a threat is that the company may define what is threatening to a person and what is not, based upon the individuals involved and the circumstances.

5.12 CONDUCTING EXIT INTERVIEWS

Conducting interviews with employees who voluntarily or involuntarily terminate employment is a valuable research tool. Conduct exit interviews with all employees who leave, for three reasons:

- To allow the employee to give the company information that can help make the company a better place for all employees to work;
- To monitor the circumstances that have caused the employee to leave. If employees that are leaving are voicing unhappiness, frustration or anger about working conditions, you may be able to take appropriate preventive actions and retain a valuable employee; and
- If an employee is leaving and makes threatening remarks, you can alert the crisis management team.

5.13 REVIEWING AND UPDATING SECURITY PROGRAMS AND PROCEDURES

The security policies and procedures you develop require regular updating. You should not develop and implement them just once and expect to have an effective Security Program. Aspects of security that require the most careful monitoring are:

- Access to facilities for employees, customers, vendors and other persons.
- Security guards and the company functions and locations for which they are responsible.
- Locations of and the need for “safe rooms”, (an office with a solid door, a lock and a telephone).
- Security devices that are designed to detect unauthorized entry to targeted locations or to alert a law enforcement agency or company security personnel of an emergency.

5.14 DOCUMENTING EVENTS

Every event that logically requires intervention, counseling, a disciplinary action or an arrest should be documented in either the employee’s personnel records or in a security report. These events may occur in person, through the mail or by means of a telephone call.

A periodic review of either personnel or security reports may reveal an escalating pattern of behavior. These reports may also be introduced into legal proceedings if a crime occurs; used to prosecute or defend against an insurance claim; and to help the crisis management team devise more effective policies and procedures.

5.15 INSTALLING EMERGENCY ALARMS

Certain areas of a building may require alarms to make employees more comfortable — particularly isolated and totally-enclosed rooms with no outside exit, ventilation, lighting, temperature controls or communication systems. Alarms are available that address a variety of needs: signal, automatic telephone dialing, intrusion, motion and light- and heat-sensing. Cash-handling facilities should always be alarmed.

5.16 INSTALLING SURVEILLANCE CAMERAS

Several closed circuit television (CCTV) cameras may be monitored from a central location, and they may be used to track movement and to identify people requesting access to a location. A master videotape may be produced from each of the selected camera locations and this videotape may help the crisis management team to identify potential errors in the company’s plan. The company may also be able to use videotapes to document an offender’s behavior and workplace violence events.

5.17 Involving Local Law Enforcement

The company's Security Director and appropriate executives and managers should interact with local law enforcement agencies periodically. The agencies appreciate the opportunity to learn about your facility in case they have to respond to an emergency there, and they may be able to provide training to your employees. Providing the police and fire departments with maps and diagrams of your facilities is also helpful.

5.18 Crime Prevention through Environmental Design: CPTED

Crime Prevention through Environmental Design is the process of designing or changing a facility in a way that controls a person's behavior, by modifying the environment that the person works in or passes through. The process emphasizes the design and use of pedestrian traffic flow, placement of furniture, lighting and other factors to control movement and to limit access. The process also addresses outside factors, including the placement and density of vegetation; access to roofs and stairwells; and forcing a vehicular traffic pattern to carefully selected and controlled areas only.

5.19 Conducting Training Programs

Developing a crisis management plan and emergency procedures is a worthless exercise unless you also train your employees how to respond effectively. An employee who is comfortable with and knowledgeable about his/her duties, authority and working environment is less likely to experience work-related problems and is more likely to respond appropriately during an emergency. All employees should be trained about basic and advanced job skills, and emergency procedures, at these times during their employment, and some of the training time should be devoted to workplace violence issues:

- At time of orientation, when you first introduce the employee to your company and your other employees.
- Whenever an employee is promoted to a new position or is transferred to another facility.
- Whenever there occurs a significant change in technology, procedure or company leadership.

5.20 Establishing and Training A Crisis Management Team

Forming a crisis management (emergency response) team is one of the most important things a company can do to reduce the potential for acts of workplace violence.

Team members should include the company's:

- Security Director.
- President/CEO.
- Director of Human Resources.

- Department heads.
- Public Affairs Representative.
- Legal counsel.
- Training Supervisors.

Supervisors often get to know employees on a personal basis. Because they are in constant contact with employees, they are often able to identify issues that could cause acts of workplace violence before those issues become problems.

5.21 Training Personnel about Conflict Resolution Techniques

Most conflicts arise because of a mistake or a misunderstanding. If not dealt with, appropriately and in a timely way, mistakes and misunderstandings become conflicts. Conflict resolution encourages the discussion and resolution of problems and discourages confrontation.

5.22 Providing Employee Safety Education Programs

Education programs should teach employees about the company's policies and procedures and how to implement them. Safety education programs teach employees about alternatives to becoming a victim and stress prevention tactics.

5.23 Providing Selected Services

People often resort to committing acts of violence when they believe that they have no alternatives and no hope. The company can provide several services to reduce the potential for acts of workplace violence, including:

5.23.1 Counseling or EAP Services
The services that people find most useful in an Employee Assistance Program are counseling for self and family members; substance abuse treatment; and stress reduction training. By emphasizing that many people participate in counseling programs to help them remain healthy, the company can often overcome the public's general belief that only "weak" people need help.

5.23.2 Outplacement Services
Outplacement services are useful for helping the furloughed or terminated employee to find other employment, on either a temporary or permanent basis. Such services show the employee that the company cares about the employee's future. This often effectively diffuses the individual's anger, frustration and potential for violence. In other words, by helping employees to find other jobs, you are lowering their desperation level.

5.23.3 Post-Trauma Services

Critical Incident Stress Debriefing", "Traumatic Events Response" and "Crisis Intervention" are terms commonly used to describe the process used to help an employee overcome the effects of a traumatic incident. Prompt treatment for a mental or emotional injury is just as important as it is for a physical injury. Both types of injuries, if they are left untreated, will continue to cause problems for the employee and the employer.

5.24 SUMMARY

Above all, company policies, procedures, and operations should be fair. Fairness means open- and even-handed enforcement of the rules. If employees know that the company is trying to be fair in the way that it operates and handles problems, there will likely be fewer incidents that cause suspicion and confrontation, which may lead to acts of violence.

6.0 WHAT SUPERVISORS CAN DO

6.1 OVERVIEW

Supervisors have a pivotal role in preventing acts of workplace violence and in handling incidents if prevention efforts fail. Supervisors generally know their employees well and can often monitor emerging conflicts that may have the potential for violence. Both business leaders and employees must be able to rely upon the knowledge, experience and judgment of the supervisor. Supervisors face these significant challenges connected with violence in the workplace:

- Developing and promoting a clear violence policy.
- Soliciting continuing support from management.
- Developing the necessary knowledge of available services.
- Maintaining continuous interaction with employees.
- Developing the ability to handle conflict.
- Following accepted violence prevention procedures.
- Promoting And Enforcing Policies.

Every company should have a clearly written and easily understood policy and step-by-step procedures for dealing with violence in the workplace. It is equally important that employees know that management is willing and able to deal with acts of workplace violence, and potentially violent situations.

The policy must describe appropriate standards of behavior for employees and other persons who have a legitimate right to be on the premises. Use the policy first as a prevention tool, and later as a disciplinary tool if your prevention efforts fail. If employees understand what the policy is and they know what the consequences are for policy violations, they are more likely to comply with the policy rather than to face a certain disciplinary action.

In addition, both employees and non-employees on the premises need to know that they have alternatives when they have a problem or a concern. If people think that they have no redress for grievances, they may become upset or angry. When people know that they have alternatives for dealing with their problems, they are less likely to become enraged.

As with every policy, a procedure must describe the method by which the policy is implemented. This involves a step-by-step analysis of the company's complaint or grievance procedure.

6.2 RECEIVING SUPPORT FROM MANAGEMENT

Executives and managers must be educated about workplace violence, and about appropriate prevention and reaction strategies. Executives and upper-level managers are often insulated from situations that happen on a day-to-day basis, and they may not recognize the true scope of the problem nor their potential liability — civilly and criminally.

Employees often feel vulnerable and unprotected when management does not deal effectively with conflict and complaints. Employees may feel as if management is ineffective; that management might be frightened of workers; or that management is uncertain of what to do before or after a violent event. A well-trained supervisor serves as a two-way communication channel between staff and management. The supervisor must inform management of employee concerns and must inform employees of management's policies.

6.3 OBTAINING INFORMATION ON AVAILABLE SERVICES

It is unrealistic to believe that a supervisor knows about all of the available resources for the prevention and reaction to violent acts. The supervisor should know what company resources exist, and where to go to get other appropriate help when it is needed. The supervisor should be aware of local resources, such as physicians and hospitals; mental health and law enforcement agencies; and the alternatives available within the judicial system. Examples are Alcoholics Anonymous (and other twelve step programs), church-affiliated programs, residential treatment programs, out-patient clinics, restraining orders and arrest procedures.

6.4 INTERACTING WITH EMPLOYEES

Supervisors manage people and work — very time-consuming tasks. They are very busy people but they should take time to interact with the people they supervise. Supervisors can spend very constructive time simply asking and answering questions, and monitoring employee relationships and behavior. Some information gathered by doing this can be a key to improving the company, and some of it can be critical to reducing the potential for acts of workplace violence.

The supervisor is the conduit for the flow of all important information within the company. By monitoring the "pulse" of the organization and by paying attention to employee feedback, the supervisor frequently becomes aware of potentially hazardous situations before they become dangerous.

Supervisors should be active and make themselves available. It is critical that the supervisor knows employees on a somewhat personal basis, because employees often tell them things that happened or are happening in their personal lives that may affect other persons in the workplace.

6.5 LEARNING HOW TO HANDLE CONFLICT

Conflict is confrontational and often unpleasant. Avoiding confrontation is natural and normal because it is stressful. One of the supervisor's primary responsibilities is to identify and resolve sources of conflict before the situation becomes volatile.

The most common sources of conflict involve the relationships between:

- People.

- Tasks.
- Schedules.
- Personal issues unrelated to employment.

We all sometimes choose to ignore the inappropriate behavior of a person because we are frightened of the person, or we are afraid that confrontation will cause a violent incident. We think — mistakenly — that ignoring the problem is better and that it will just go away. History indicates that this just does not work.

If we ignore inappropriate behavior, we are condoning it — and often encouraging it, as well. Taking action to handle conflict can be risky and it involves setting a good example, promoting a "zero tolerance" violence policy, addressing employee concerns about confidentiality in reporting threats, and following a violence prevention procedure.

6.6 SETTING A GOOD EXAMPLE

We tend to adopt the qualities and attributes of the people we live and work around, and we all lead by example — positively or negatively. The supervisor's attitude and behavior creates the example for others to follow. If you are often annoyed, constantly criticize management's decisions and are discourteous to others, your employees will be, too. The supervisor must become the positive role model and counsel those employees who display negative behavior.

6.7 PROMOTING A "ZERO TOLERANCE" VIOLENCE POLICY

We all have to deal with people who have either assertive or aggressive personalities. Assertive here means to strive for what we want while respecting the rights of other persons. Aggressive here means insisting on getting what we want, often at the expense of other persons. Aggression often leads to confrontation if the aggressor is frustrated in his/her attempts to get what he/she wants. The supervisor must not tolerate mental, emotional or physical aggression and must counsel those employees who display aggressive behavior.

6.8 ADDRESSING EMPLOYEE CONCERNS ABOUT CONFIDENTIALITY IN REPORTING THREATS

It is normal to dislike and distrust people who "snitch" or "tell on" others. We also often perceive a person as "weak" who cannot or will not handle his/her own problems. We hire people as employees to perform work, however, and not to have to defend themselves from offenders in the workplace.

Employees have a right to work within a safe and secure environment, and also to be able to report their workplace problems and concerns in a confidential manner. The supervisor must treat all such reports as confidential and should notify the Security Director of issues that he/she cannot handle alone.

6.9 FOLLOWING A VIOLENCE PREVENTION PROCEDURE

The company should establish a well-planned procedure for preventing acts of violence. A violence prevention procedure is intended to identify and address problems before they result in more serious behavior, such as threats or violent acts. A violence prevention procedure includes:

- Observing employees and events.
- Documenting performance and events.
- Initiating employee meetings.
- Making appropriate referrals.

6.10 OBSERVING EMPLOYEES AND EVENTS

Supervisors must know how to manage and monitor employees — to promote good work and to correct mistakes. Supervisors should also be watching for "at risk" employees that may be experiencing frustration or difficulty that affects other people in the workplace environment.

6.11 DOCUMENTING PERFORMANCE AND EVENTS

Supervisors must document positive and negative performance and behavior changes, events and real or suspected violent acts for legal, psychological and human resource reasons. The documentation process should begin when the supervisor first observes a person's obvious change in attitude, behavior or performance.

6.12 INITIATING EMPLOYEE MEETINGS

Employee meetings are an excellent way to train, motivate and monitor employees, since all outside distractions are absent. It is just the supervisor and the employee — one-on-one. If you arrange such a meeting to discuss a current or potential problem, follow these guidelines when you approach the event you want to discuss:

- Schedule the meeting at a private location and at a time that is mutually convenient.
- Arrange to have an impartial witness present during the meeting if you believe it is appropriate.
- Express your concerns, describe the behavior and give examples.
- Focus upon the behavior causing or contributing to the event.
- Do not discuss the potential impact yet.
- Do not take notes.
- Ask the employee for his/her viewpoint and comments.
- Listen and encourage discussion.

- Summarize what the employee said.
- Tell your point of view.
- Allow the employee to respond.
- Summarize what the employee said.
- Discuss the potential impact now.
- Ask the employee for his/her input;
- Ask the employee to think how he/she will approach the problem.
- Set specific guidelines and a realistic timetable for change.
- Express confidence and support for the planned changes.
- Schedule a follow up meeting with the employee to review his/her progress, before you bring this meeting to a close.

6.13 Making Appropriate Referrals

Obvious changes in a person's behavior and productivity on the job may indicate that the individual is experiencing something unusual or unfortunate elsewhere in life. Anyone may experience emotional situations occasionally, but a dramatic change in behavior generally indicates that the supervisor should intervene. Referral to the company's Employee Assistance Program may be helpful.

6.14 Managing Threatening Behavior

When a person's anxiety, fear or frustration reach's a critical proportions, he/she may threaten another person. A threat is simply a warning that something is not "right" in the person's world and that he/she plans to do something about it. A threat may be used to disguise a person's true feelings; to intimidate another person; or to send a message about the person's intention to act. Handling threats effectively requires special communication skills — and a plan.

6.14.1 Responding To Threats

Threats can be implied or expressed. An implied threat causes a person to feel threatened without any real evidence of a threat being made. An expressed threat is a deliberate remark made or action created that leaves no doubt about the offender's intentions.

6.14.2 Handling Threats

If the supervisor receives or learns of a threat, he/she may first want to secure consultation before proceeding. Then, if it is appropriate, the supervisor should assess the seriousness of the threat, and establish the context of the threat in a planned, strategic way.

6.14.3 Securing Consultation

After you have heard the threat, secure consultation. Do not try to handle it by yourself — and do not ignore it. Believe the threat and take it seriously — then get assistance. For most companies, assistance may be available from the Security Department and/or the Human Resources Department. Security Directors and human resource professionals maintain confidentially. They can talk to you about the threat and they will have background information relating to the person in question. Remember — a person's history is key to determining their potential for violence.

6.14.4 Establishing Context

Establish the context of the threat by asking other employees — and yourself — several questions, including:

Is this person emotionally upset (e.g., argumentative, inattentive, accident-prone)?

- Has this person been abusive with or abused by other employees (e.g., confrontational, used profane language or made threatening or obscene gestures)?
- Has this person displayed recent and significant changes in work habits (e.g., tardiness, incomplete work or unkempt appearance)?
- Has a significant change occurred in this person's life that could be responsible for the threatening behavior (e.g., recent family illness or death, financial adversity or an unstable marriage)?
- Could this person be experiencing continuing problems in his/her personal life (e.g., drug or alcohol addiction, bankruptcy or a divorce)? and
- Has this person's professional and career goals stalled (e.g., failure to pass a promotional exam, unsatisfactory performance review or a demotion)?

6.15 Taking Appropriate Action

The actions a supervisor must take to resolve threats are different from actions taken to facilitate intervention. These actions are classified as non-emergency and emergency responses.

Non-Emergency Responses

The supervisor should contact the company's Security Department and the managers who are members of the company's Crisis Management Team immediately. The Security Department have members of the company's Guard Force remove the intended victim or the threatening employee from the premises, and it may also secure the immediate area from trespass by people and vehicles. The Crisis Management Team may also require that other actions should be taken on an interim basis.

Emergency Responses

The supervisor should contact the nearest law enforcement agency and the company's Guard Force immediately if danger is imminent, and then contact the company's Security Department. The supervisor should advise the agency about the threat, the persons and locations involved, and the circumstances of the emergency. The law enforcement agency will respond with the types and numbers of personnel that it considers appropriate.

6.16 SUMMARY

The supervisor is a supervisor because of above-average administrative and operations skills. Preventing workplace violence may seem like just another responsibility to be heaped on the pile, but it is really not one. A well-trained supervisor already has the people skills and communication skills to help prevent workplace violence. The only thing lacking may be the awareness of what constitutes a potential problem and the appropriate solutions. Remember — take all threats seriously; secure consultation; establish context; and react appropriately. If you do not do something about workplace violence, it may very well do something to you.

Notes

7.0 WHAT SECURITY AND LAW ENFORCEMENT CAN DO

7.1 OVERVIEW

The Security Director is responsible for providing a safe and secure environment for employees and customers to interact. The Security Director must have the authority to design and implement an effective Security Program, and the authority to use any and all available company resources during an emergency.

A company generally has no responsibility for protecting an employee while he/she is away from the workplace and not involved in company-related business. However, a company should initiate special security measures when an employee brings potential problems to its attention — problems that may affect the workplace. If it does not, the company may be held responsible for its inattention, inaction or neglect.

According to the current "reasonable and prudent care" standard, a company may be held liable if it does not have an adequate Safety and Security Program in place — if a violent event occurs. This standard also requires that the company must be able to investigate all threats of violence in a prompt and effective way, and then take action to prevent the act from occurring — or at least to reduce the potential effects of the act. In order to meet this standard, employees should be trained about the company's Security Program so that they know:

- Who is in charge.
- What needs to be done.
- How to do it.

7.2 WHAT PRO-ACTIVE SECURITY CAN DO

Being "pro-active" means anticipating a problem and preparing for a crisis before it happens. Being "reactive" means developing a response plan —during or immediately following the event — that requires an emergency response. You can become pro-active by:

7.3 FORMING A SECURITY COMMITTEE

The company should first appoint appropriate staff, supervisor, middle management and executives employees to a Security Committee. The Security Committee should be responsible for insuring that the company's employees, customers, non-affiliated third parties, facilities, assets and records are protected against internal and external threats.

The Security Committee should also be responsible for selecting a Security Director who directs the company's Security Program and establishes the company's Crisis Management Team. The Crisis Management Team should contain the best leaders and managers the company has to offer and seniority should not be the primary qualification.

7.4 DESIGNATING A SECURITY DIRECTOR

The Security Director should be responsible for researching, developing and implementing the company's Security Program, and for managing the company's Guard Force. The Security Director should report directly to and should be a member of the Security Committee. The Security Director should also be responsible for reporting — at least annually — on events and conditions that may affect the safety and security of employees and other company functions.

7.5 DEVELOPING A SECURITY PROGRAM

The company's Security Program should contain policies and procedures designed to reduce risks from any threat, and to assist employees in responding to potentially hazardous events. A Security Program is evolutionary and requires periodic updating to remain effective. This periodic updating is based on the results of the testing and training programs specified in the Security Program and should require the participation of all employees.

7.6 CONDUCTING A RISK ASSESSMENT

You cannot develop and implement an effective Security Program without first identifying and prioritizing potential risks. Depending upon your risk assessment, consider security steps such as:

- Illuminating walkways, parking areas and other isolated locations;
- Providing a clear view for employees to observe exterior areas, and for law enforcement agents to check interior areas of your building;
- Providing enclosed private parking for employees that is not accessible by non-employees;
- Using access-control techniques and equipment, such as sign-in logs, door controls, card or digitized access devices, employee- pictured identification and color-coded visitor badges;
- Installing silent alarms wired to a security alarm company or your local law enforcement agency;
- Developing internal distress or warning codes, emergency procedures and a visual and/or verbal means of alerting employees to a dangerous incident within the building;
- Using security guards during peak crime-time hours for targeted prevention and escort services;
- Conducting pre-employment investigations on all applicants; and
- Issuing bulletins for employee briefings about offenders, potential problems and hazards.

7.7 PRACTICING CRIME PREVENTION THROUGH ENVIRONMENTAL DESIGN: CPTED

The Security Department may make safety and security recommendations based on Crime Prevention Through Environmental Design — or CPTED — standards. The concept behind CPTED is to control a person's behavior by changing and modifying the physical work environment.

Examples of environmental design solutions include the placement of desks, counters and partitions; installing warning signs and bullet-resistant barriers; placing cameras and alarms in strategic locations; and replacing metal locks with key-card access devices.

7.8 DEVELOPING PREVENTION AND RESPONSE STRATEGIES AND TACTICS

Recognize that the first (and often a continual) step in this process of creating a safe and secure environment is to educate yourself and your employees about workplace violence, using the time-honored standard of who, what, where, when, why and how. You have to know your adversary before you can develop a winning strategy. Your company can then do many things simultaneously to reduce the opportunity for your employees and customers to become the potential victims of violent crimes by determining:

- **WHO**: Who is most at risk to these crimes? (e.g., staff, supervisors, managers, executives, directors — and their families, or customers and vendors).
- **WHO:** Who is most likely to commit these crimes? (e.g., transients, area residents, professional criminals, gangs, family members, former employees or dissatisfied customers).
- **WHAT**: What crimes are most likely to occur? (e.g., create projections based upon the geographic area, industry and company history and significant changes in surrounding social and cultural structures).
- **WHERE**: Where are the most likely sites for specific crimes to occur? (e.g., robbers often prefer ATMs and convenience stores located in metropolitan areas for their crimes).
- **WHEN**: When are specific crimes likely to occur? (e.g., most crimes affecting cash-handling functions of retail companies are committed during peak customer traffic times, when distractions may create confusion among customers and a lack of focus among employees).
- **WHY**: Why are specific crimes committed? (e.g., burglars burglarize to steal property, robbers rob for money, terrorists terrorize to draw attention to a personal or political cause); and
- **HOW:** How are specific crimes committed? (e.g., robbers use verbal commands and gestures, notes, firearms, bombs and other weapons.)

7.9 DEVELOPING A DOCUMENTATION PROCEDURE

The Security Director must become aware of everyday circumstances that may pose a risk to the company. The Crisis Management Team must be made aware of unusual events that require more attention than the Security Director alone can give. Developing and implementing a documentation policy and a procedure — for relaying the necessary information to the Security Director in a timely manner — is critical.

7.10 DEVELOPING A COMPANY-WIDE TRAINING PROGRAM

Because coping with and surviving violent incidents is often a team exercise, the success of the Security Program depends upon training employees — at all levels — to function as a cohesive unit. Employees benefit from standardized training when they work at their normal locations, and they benefit when they work at different locations.

7.11 COORDINATING YOUR PLAN WITH LOCAL LAW ENFORCEMENT AGENCIES

If the Security Department anticipates an event or is advised of an event in progress, it may be able to coordinate emergency responses among other departments and law enforcement.

7.12 MAINTAINING LIAISON WITH LAW ENFORCEMENT

The Security Department is responsible for assisting local law enforcement officers. This assistance may be limited to holding training meetings; providing information, providing company resources and personnel; containing the event or person; or simply to get employees out of the way.

7.13 PRACTICING EFFECTIVE INTER-DEPARTMENT COMMUNICATION

The Security Department often serves as a conduit — a clearing house and a central source — for information, particularly of a sensitive nature. Encouraging employees to observe and report unusual or suspicious incidents and crimes directly to the Security Department establishes a communications link that may be used to the company's advantage. If employees have not been told where to report these events, they likely will not report them at all.

7.14 SHARING INFORMATION

When a potential security problem is just starting to develop, it is often being addressed by more than one department. It is essential that each department understands what the other department is doing and coordinates their information-gathering and problem-resolution efforts with the Security Department.

7.15 NOTIFYING APPROPRIATE PERSONS

The Security Department is often aware of a potential threat before other departments know about it — and sometimes even before the target of the threat is aware of it. The Security Department should be able to operate confidentially and to have the opportunity to make the appropriate notifications quickly and quietly.

7.16 CONDUCTING PRE-EMPLOYMENT AND INCIDENT INVESTIGATIONS

The Security Department should participate in or conduct pre-employment background investigations, and also should conduct investigations regarding an employee's conduct after the employee comes to work.

7.17 INVESTIGATING THREATENING CALLS

The Security Department may install equipment to monitor telephone calls, and to record them if necessary. In some states, telephone calls, recorded with the permission of just one party, are admissible in court.

7.18 INVESTIGATING THREATENING BEHAVIOR

Anyone may display offensive and potentially threatening behavior in the workplace. The Security Department's Guard Force may be responsible for contacting the offender and de-escalating the situation. They may also have the responsibility for escorting the offender from the premises or arranging for the offender's arrest.

7.19 CONDUCTING INCIDENT DE-BRIEFING MEETINGS

We make our best decisions when we are fully informed. A de-briefing allows the Security Director to present a synopsis of the event; to discuss effective and ineffective responses by company personnel and other persons involved in the incident; and to seek suggestions for alternative responses is a similar incident occurs in the future.

The de-briefing also allows the Security Director to discuss areas of concern that will logically arise after the incident becomes public knowledge, including:

- Changes to inter-departmental communications and the reporting structure.
- Anticipated changes in company policy or procedures.
- Suggestions about how employees should discuss the incident among themselves and with their families.
- The company's policy regarding talking with customers or with media representatives about the incident.

7.20 PROVIDING ESCORT ASSISTANCE

The Security Department's Guard Force should provide an escort service to help employees get to their cars safely. This service should be made available before, during and after work hours — or whenever an employee requests it. The company may also make the services of a taxi company available for employees who work late.

7.21 WHAT LAW ENFORCEMENT CAN DO

Law enforcement agencies, like their civilian counterparts in security organizations, prefer to deal with problems in the workplace before they escalate into violence. Law enforcement agencies often offer pro-active services such as:

- Liaison officers.
- Site surveys.
- Referrals.
- Violence assessments.

7.22 LIAISON OFFICERS

Also known as community services officers, liaison officers concentrate their efforts on preventing crimes instead of apprehending offenders. Liaison officers develop and share crime prevention information among other agencies and with local businesses. They are often one of the best — and most readily-available — Security Department resources. The liaison officers also have access to confidential information about potential offenders. Although some criminal records are confidential, certain background information may be shared if an existing or emerging situation demands the "need to know".

7.23 SITE SURVEYS

Liaison officers are frequently members of a crime prevention or community policing unit. They can come to your organization, evaluate your security procedures and recommend added or modified security measures. Often this is a free service that local law enforcement agencies provide. Part of this service may involve conducting a site survey. Implementing site survey recommendations may prevent breaches of security. Common recommendations involve:

- The placement of unobtrusive barriers.
- Installing specialized locks and tamper-resistant windows.
- Adding or modifying lighting, camera and alarm devices.
- Creating security-conscious premises and grounds keeping designs.

7.24 VIOLENCE AND PREMISES ASSESSMENTS

Law enforcement agencies may perform violence and premises assessments in anticipation of a violent crime. If the officers have to respond to an emergency at your company, they want the most current information available about the

company's personnel, layout, operations and history. Law enforcement agencies may also provide the company with referrals to private security firms for specialized assessment services.

7.25 NON-VIOLENT INTERVENTION

In the interest of public safety, law enforcement officers may be able to contact an offender, even if no crime has been committed. Generally all the officers need is a call to request their presence. The officers can warn the potential offender of laws related to criminal trespass, public disturbance and violating a restraining order, if one exists.

7.26 REFERRALS

Law enforcement officers are trained to de-escalate potentially violent situations, and to know when arresting an offender is the most appropriate solution. Arresting a person may not always be the best course of action. It may simply make a bad situation, worse — and it might otherwise have taken care of itself.

Officers often use both their obvious authority and implied power to refer offenders to formal and informal sources of help. The formal sources may be treatment programs or departments within the criminal justice system. The informal sources may be friends, family members or mental health practitioners who volunteer their services.

7.27 WHAT TO DO NEXT

Creating appropriate prevention techniques requires that your company acknowledges the potential for workplace violence; prepares strategies to address issues likely to occur; and trains its employees to behave appropriately during an event if prevention tactics do not work. You can take many pro-active steps to reduce the opportunity for an act of workplace violence to occur. Develop your prevention strategy using the information you gather by following a few simple guidelines, including:

- <u>Investigate</u>

  Research the problems experienced by other companies — particularly similar companies — and determine what techniques they have used successfully to reduce risks. Consider starting or participating in a local crime prevention group.

- <u>Identify</u>

  Examine each worksite — not just cash-handling facilities — and the areas surrounding those worksites for conditions that may actually encourage crime (e.g., poor lighting, inadequate visibility or isolated locations).

- Assess

  Review your operating policies, procedures, and training programs for misleading or incorrect information about what employees should do during an emergency.

- Communicate

  Create a communication policy and follow up procedures that employees are encouraged to follow when reporting possible risks. The policy should require that employees report threats or acts of intimidation to the Security Director that originate either at home or at work

- Train

  Invite local law enforcement and other agencies to present training programs about domestic violence, sexual assault and other forms of personal assault prevention techniques. This training may help employees to recognize the characteristics of offenders, the events that may occur, and appropriate prevention or resolution techniques.

7.28 SUMMARY

You should think of private security and public law enforcement as partners in crime prevention. Working closely with security and law enforcement personnel before an incident occurs is preferable to being forced to react to an incident that is already out of control. Following well-designed procedures can familiarize employees with security and police personnel and help the company to avoid traumatic acts of workplace violence.

8.0 POLICY AND PROCEDURES

8.1 WORKPLACE VIOLENCE PROCEDURES: ALL PERSONNEL

8.1.1 PURPOSE: The purpose of this section is to define and describe the company's policies and procedures regarding workplace violence. The objective is to ensure the highest standards of health and safety for all employees, customers and other persons who may be present, and to provide for the efficient and effective operation of the company.

8.1.2 No employee or non-employee present on the company's premises will be allowed to harass any other employee or non-employee. Any person who violates this policy is subject to immediate disciplinary action, and may also be subject to arrest and prosecution by a law enforcement agency. Examples of prohibited behavior include, but are not limited to:

1. Verbal harassment: Making verbal threats toward persons or property; using vulgar or profane language; making disparaging, derogatory or insulting remarks; initiating offensive or unsolicited physical contact, sexual flirtation or advances; and any other act of verbal intimidation.
2. Physical harassment: Initiating any physical assault such as hitting, kicking, pushing, holding, impeding or blocking the movement of another person. No person will forfeit the basic right to self-defense, however.
3. Visual harassment: Making or displaying disparaging, derogatory or insulting posters, cartoons, drawing, slogans or publications.
4. Prohibited items on company premises: Possessing or displaying firearms of any kind; possessing or displaying switchblade knives or other knives with a blade longer than four inches; possessing dangerous chemicals or explosives, including blasting caps; and possessing any other objects, lawful or unlawful, that are carried for the purpose of intimidating or injuring another person.

8.1.3 All employees will assist in preventing acts of workplace violence by:

1. Notifying the Security Director and/or the most readily available supervisor of any unusual actions displayed by a current or former employee, including any person who:
    1. Constantly complains about the workplace environment in general;
    2. Talks about a real or imagined grievance against a supervisor, or who makes verbal threats or physically confronts the supervisor;
    3. Appears moody, depressed or unbalanced;
    4. Has a spouse or other significant person who calls or appears at the workplace and creates a disturbance; and
    5. Visits the company after being terminated, to talk with former coworkers about receiving unfair or unjust treatment.
2. Reporting any threatening telephone calls, no matter how trivial they may seem.

3. Reporting threatening incidents as they occur to the Security Director or the most readily available supervisor immediately.
4. Filing copies of personal restraining orders with the Security Director.
5. Participating in any investigation that results from the incident.

8.1.4 The Personnel Office will inform the Security Director about any employee who is disciplined, resigns or is terminated, and who exhibits any violent tendencies or who expresses an extreme dislike for a supervisor and/or coworker. The Security Director:

1. Will prepare a memorandum for the Guard Force Commander. The memo will describe the potential problem; provide a photograph and the physical description of the person, if one is available; and the names of possible targets.
2. Will conduct an investigation of the incident and notify members of the company's Crisis Management Team of the result.

8.1.5 The company's Crisis Management Team consists of at least a(n) Vice President, Security Director, Director of Human Resources, Administrative Officer and Guard Force Commander. If an act of violence occurs, the Crisis Management Team will assemble and:

1. Notify the nearest law enforcement agency of the event;
2. Establish a Command Post at one of the following locations:
    1. Guard Force office.
    2. Security Director's office.
    3. Conference room.
    4. Executive Suite;
    3. Establish liaison with police.
    4. Establish communications.
    5. Interview, if possible, the person(s) who knows the offender and may be aware of his/her motivation.
    6. Notify the law enforcement agency if the offender has a hostage, either the intended victim or another person.
    7. Assist the law enforcement agency's negotiator by providing cooperation and access to the company's facilities.

8.1.6 If an act of violence occurs, the Guard Force becomes responsible for:

1. Containing the offender within an established perimeter, if possible;
2. Establishing traffic control on company property;
3. Escorting employees who may be of assistance to the Command Post;
4. Evacuating appropriate buildings and adjacent areas; and
5. Obtaining standby medical assistance.

8.1.7 When the incident is terminated, the Crisis Management Team will decide whether to return personnel to their respective offices immediately or to wait for other instructions.

8.2 CRISIS MANAGEMENT TEAM: DUTIES AND RESPONSIBILITIES

The duties and responsibilities of the Crisis Management Team include all phases of the Crisis Management Plan:

1. Delegating the initial and continuing research and development of specific elements of Crisis Management Plan to appropriate personnel, as those elements pertain to their departments and/or assignments.
2. Ensuring that appropriate guidelines are provided for the initial and continuing research, development and implementation of the Crisis Management Plan.
3. Reviewing the information gathered; create or update each office, departmental and function plan for inclusion into the total Crisis Management Plan.
4. Arranging for the initial and continuing implementation and testing procedures involved with all phases of the Crisis Management Plan.
5. Acting as liaison with law enforcement, fire and medical personnel, and providing the Security Director with timely and complete updated information regarding any emergency addressed by the Crisis Management Plan.
6. Ensuring that Incident Logs are prepared by all departments, indicating events and measures taken to address the impact of the emergency.
7. Ensuring that their specific areas of responsibility are staffed with appropriate personnel, and that these personnel are working in a safe environment.
8. Coordinating the post-emergency evaluation effort, and advising the Security Director of the results by means of a written report.
9. Participating in the company's periodic and company-wide evaluation of the Crisis Management Plan and providing written recommendations regarding changes to be made.
10. Scheduling and holding at least annual meetings with all Crisis Management Team members, and ensuring that each meeting is documented regarding:
    1. The purpose of the meeting.
    2. Those persons in attendance.
    3. Agenda items addressed.
    4. Revisions proposed during the meeting that are to be discussed and approved during the next regularly-scheduled Board of Directors' meeting.

8.3 SECURITY DIRECTOR: DUTIES AND RESPONSIBILITIES

The Security Director is responsible for the following duties and responsibilities:

1. Ensuring the timely, continual and effective research, development and implementation of the Crisis Management Plan.
2. Developing, assigning and monitoring the duties and responsibilities of the other Crisis Management Team members.
3. Reviewing the results of each developmental, implementation and testing phase of the Crisis Management Plan when appropriate, and arranging for the plan's annual certification by the company's Board of Directors.
4. Maintaining a current master copy of the Crisis Management Plan secured at his/her primary workplace, and a duplicate master copy at his/her place of residence or another appropriate location.
5. Upon notification of a potential or current emergency, determining the potential and/or current damage to persons, facilities and company operations.
6. Verifying that all appropriate law enforcement, fire and medical services have been notified of the emergency.
7. Ensuring that all Crisis Management Team members have been notified of the emergency, either in person or by phone, and that those members are responding immediately to the Crisis Management Center of Operations.
8. Determining the cause of the emergency, if possible.
9. Ensuring the execution of necessary measures to provide for the safety of personnel, customers and other persons on the premises, facilities, assets and records.
10. Initiating the shift in operations to the designated Crisis Management Center of Operations, if appropriate.
11. Providing the Board of Directors and other appropriate personnel with updated information on a timely basis, regarding this emergency and its impact.
12. When the emergency has ended, ensuring that all measures taken by the company's personnel during the emergency are recorded.
13. Presenting a synopsis of the results to the Board of Directors for evaluation.
14. Ensuring that the company's Crisis Management Plan was implemented as appropriate, and/or arrange for the plan to be modified based upon the evaluation of this emergency response.

8.4 DEPARTMENT LEADERS: DUTIES AND RESPONSIBILITIES

The Crisis Management Team members will appoint Department Leaders (Office Managers will function as Department Leaders), who will be responsible for the following duties and responsibilities relating to the Crisis Management Plan:

1. With the assistance of the Crisis Management Team members, researching, developing, implementing and continually monitoring a comprehensive emergency response and crisis management policy and procedure program. This program will integrate existing precautionary programs with programs yet to be developed.
2. Maintaining files for his/her department and facility regarding emergency response and crisis management procedures to be performed by that department or facility.
3. With the assistance of the Crisis Management Team members, researching, developing, implementing and monitoring an emergency evacuation plan for his/her department or facility, and submitting each plan, and at least annual evaluations of the plan's effectiveness, to the Crisis Management Team members for review.
4. With the assistance of the Crisis Management Team Coordinators, designating and training emergency leaders and assistants for all his/her department or facility, to assume responsibilities during emergency responses.
5. Researching, developing, implementing and monitoring a comprehensive emergency response and crisis management training program for all personnel, according to the guidelines supplied by the Crisis Management Team members.
6. Maintaining a current master copy of the Crisis Management Plan secured at his/her primary workplace, and a duplicate master copy at his/her place of residence or another appropriate location.
7. Recommending to the Crisis Management Team members the acquisition of all sites, equipment, supplies and services necessary for his/her department or facility to respond adequately to an emergency.
8. With the guidance of the Crisis Management Team members, managing and controlling all aspects of emergency activation and crisis management planning, testing, implementation and maintenance of systems, procedures and policies for his/her department or facility.
9. Collecting, analyzing and reporting upon the facts concerning all departmental functions relative to emergency activation and crisis management for his/her department or facility.

10. Conducting at least annual reviews and assessments of the protective measures designed to reduce risks, and furnish those written results to the Crisis Management Team members.
11. At the direction of the Crisis Management Team members, implementing crisis management procedures and the conversion to normal operations when the emergency has ended.
12. Providing the Crisis Management Team members and other appropriate personnel with updated information on a timely basis, regarding this emergency and its impact.
13. Ensuring that all measures taken by the company's personnel during the emergency are recorded, and presenting the recorded results to the Crisis Management Team members for evaluation.

8.5 CRISIS MANAGEMENT PLAN

If an event occurs that may require the activation of the company's Crisis Management Plan, the discovering employee will promptly notify his/her immediate supervisor of the event and its circumstances. If the immediate supervisor determines that the activation of the Crisis Management Plan is required, he/she will promptly:

- Notify the Security Director or another member of the Crisis Management Team, if the event is discovered during business hours; or
- Notify the Security Director or another member of the Crisis Management Team at the selected emergency response telephone number, if the event is discovered during non-business hours.
- Persons at the selected emergency response telephone number maintain emergency telephone numbers for Crisis Management Team members and other appropriate persons.
- Persons at the selected emergency response telephone number will initiate appropriate emergency notifications, and will also initiate an Incident Log to track events relating to their roles in the response effort.
- Upon notification by either the reporting supervisor, the Security Director or another member, the Crisis Management Team will formally initiate the Crisis Management Plan and authorize appropriate emergency responses. Selected functions may be moved to the designated Centers of Operations (Command Centers):
- Crisis Management Team members and other appropriate managers will initiate independent Incident Logs, to track events and their respective roles during response efforts.

- The Crisis Management Team will notify available members of the company's Board of Directors, at a time and in a manner the Team considers appropriate.
- The Crisis Management Team will take direction from local emergency services personnel at the scene, if possible.
- The Crisis Management Team will assemble at the designated Command Centers as assigned:
    - Team members will gather data for the initial damage assessment. Based upon this initial assessment, the Crisis Management Team will determine the appropriate level and method of response.
    - Company personnel will be assigned to emergency telephone staffing responsibilities. All personnel assigned to staffing emergency telephones will:
        - Assemble at the designated sites as assigned.
        - Initiate independent Incident Logs to track events and their respective roles during recovery efforts.
        - Maintain designated emergency:
            - Lines for contact with Crisis Management Team members and appropriate managers.
            - Lines for employees' use.
            - Lines for routine business and other business functions as necessary.
- The Crisis Management Team will manage responses by personnel, vendors and managers from the designated Command Centers. Managers will manage on-site responses by the local emergency services agencies, and other on-site functions as assigned by the Crisis Management Team, including:
    - If only facilities or assets have been damaged, representatives from the facilities management, insurance and security functions will be dispatched for on-site damage assessment and control.
    - If records also have been damaged or compromised, a legal representative will be dispatched for on-site damage assessment and control.
    - If the Data Processing facility has been damaged or is inaccessible, appropriate functions will be shifted to the designated recovery site(s).
    - If records also have been damaged or compromised, a legal representative will be dispatched for damage assessment and control.
- Based upon the on-site damage assessment, the Crisis Management Team will determine whether to:
- Close the facility, office or department, and temporarily reassign employees and customer service functions;

- Open the facility, office or department on limited basis, and advise personnel and customers; or
- Open the facility, office or department on full-service basis.
- Based upon the on-site damage assessment and the course of action determined by the Crisis Management Team, the Security Director or another appropriate person will initiate a press conference and issue a press release. This press release will include statements regarding:
    - An initial damage assessment to facilities only — statements regarding injuries, fatalities, and general logistical conditions will be referred to the responsible emergency services agencies;
    - The name(s) of the person(s) responsible for managing the company's response effort;
    - The names of company representatives authorized to release information to the media;
    - If credit functions are affected, that all customer accounts are safe and protected, and when the credits to those accounts will be available; and
- The location designated as a temporary customer service center, including:
    - The name of the on-site representative(s).
    - The physical address of the location.
    - Any telephone "hotline" number designated for customer inquiries.
    - The hours of operation.
- The Crisis Management Team will decide when to shift company resources from recovery (stabilization) to restoration (of critical functions only), resumption (of all functions), or reconstruction (to the point of incident). This includes:
    - The notification of any appropriate regulatory agencies and insurance carriers of the event and a summary of actions taken, if that is appropriate; and
    - The issuance of periodic press releases by the Crisis Management Team as considered appropriate.
- Based upon the continuing damage assessment, progress reports and recovery efforts, the Crisis Management Team will determine when to deactivate the emergency response status. When the shift to routine operations is initiated:
    - Interim policies and procedures will be implemented as appropriate.
    - Continuing restoration efforts will be delegated to the on-site representative(s).
    - Emergency telephone personnel will be assigned to contact appropriate managers and other persons at the selected emergency response telephone number, with deactivation instructions.

- The Crisis Management Team will conduct its own debriefing and critique immediately following the plan's deactivation, and:
  - The Crisis Management Team members will issue a press release, a summary of actions taken and the current response status.
  - Notify any appropriate regulatory agencies and insurance carriers of the return to routine operations, or continued restricted activities, if that is appropriate.
  - Advise the company's Board of Directors of the current response status.
- Appropriate managers will furnish a written after-action report within (24) twenty-four hours of the Crisis Management Plan's deactivation.
- Completed Incident Logs will be delivered to the Command Center, as attachments to the managers' written after-action reports.

- Within (24) twenty-four hours of the Crisis Management Plan's deactivation, the company's Auditor will evaluate the company's emergency response and:
  - Meet with the Crisis Management Team and discuss his/her findings.
  - Make recommendations to revise the Crisis Management Plan, if necessary.
  - With (24) twenty-four hours of the Auditor's review, the Crisis Management Team will:
    - Notify any appropriate regulatory agencies and insurance carriers of the company's continuing recovery efforts and a summary of actions taken, if that is appropriate; and
    - Prepare and distribute a written summary of the event and the company's recovery efforts to any appropriate regulatory agencies, insurance carriers and appropriate personnel, including the Board of Directors.

8.6 BOMB THREAT PROCEDURES: MANAGEMENT

8.6.1 PURPOSE: The purpose of this section is to define and describe acceptable management procedures for handling a bomb threat, to avoid confusion and to monitor the performance of necessary tasks assigned to staff personnel.

8.6.2 Before receiving a bomb threat:

1. Insure that appropriate opening and closing procedures are performed.
2. Encourage all employees to take appropriate actions regarding suspicious persons or events, and to report them promptly.
3. Act upon reports of unusual behavior or incidents.
4. Conduct frequent bomb threat and evacuation procedures training during staff meetings and other appropriate occasions.
5. Conduct frequent checks of workstations to insure that the stations and surrounding areas are free from debris and other items that could conceal a bomb.
6. Insure that forms assigning employees to specific responsibilities are current and complete, and that each workstation contains adequate emergency forms. Also verify that each employee knows what is expected from him/her during such events.
7. Establish strict control procedures for the inspection of packages and other materials entering critical areas.
8. Establish and maintain a tracking system for identifying and tracking employees and other persons having access to critical areas.
9. Instruct all personnel to periodically inspect rest rooms, closets, elevators, stairwells and other areas open to the public for unusual objects.
10. Organize and train search teams and evacuation teams using personnel who are familiar with the building and its contents.
11. Contact the local law enforcement agency and determine if a bomb disposal unit is available and under what circumstances this unit will respond in case of a bomb threat.

8.6.3 <u>During</u> the bomb threat:

1. If you are notified of the emergency during the bomb threat telephone call, assign the employee receiving the call to remain in contact with the caller, if possible.
2. Immediately notify the Unit Security Officer, the local law enforcement agency, the local fire department and the Security Director of the call by using another telephone line.
3. Avoid any actions that may alert customers or other persons in the office that a potential emergency situation exists.

4. Dedicate specific telephone lines as emergency lines and station an employee with those telephones to answer them if necessary, and to insure that the lines are not used for any other purpose until the incident has been terminated.
5. If the facility is so equipped, begin office closing operations immediately, moving all appropriate items to the vault.

8.6.4 After the threat:

1. If the caller states that a bomb has been placed in or around the office, a management team consisting of the Unit Security Officer, a supervisor and the Security Director will evaluate the threat level to determine if the office is to be evacuated. The Security Director has the final authority to call for an evacuation, and if any of the three individuals are absent during the threat, then the remaining members will make the decision.
2. The office will be searched whether or not it is evacuated.
3. The Unit Security Officer will advise all employees to visually search the areas immediately surrounding their workstations and to report objects that are suspicious. Any employee locating a suspicious object is responsible for reporting that object and its location only: no employee will otherwise disturb the object.
4. If an evacuation is required, the evacuation team members will evacuate all employees, customers and other persons not assigned to search procedures, if applicable.
5. If the office is to be evacuated, the Unit Security Officer will insure that all sensitive information, currency and negotiable documents are transported to the vault, if the facility is so equipped, and that the vault door is closed until the incident is terminated.
6. If the office is evacuated, all persons leaving the building will be guided to the appropriate emergency staging area(s). Wherever possible, witnesses are to be identified and asked to remain at the emergency staging area(s), pending contact by the law enforcement agency.
7. Those personnel responsible for searching the building will report their findings to the Unit Security Officer as they leave the building, and these persons will join the other employees, customers and other persons assembled at the emergency staging area(s).
8. The building will not be re-entered until it has been inspected and cleared by the responsible fire department and the law enforcement agency, and the Security Director authorizes that action.
9. All employees will remain at the emergency staging area(s) until they are authorized to leave by the Unit Security Officer or the Security Director.
10. All media inquiries will be handled by the Security Director.

8.7 BOMB THREAT PROCEDURES: STAFF

8.7.1 PURPOSE: The purpose of this section is to define and describe acceptable staff procedures for handling a bomb threat, to avoid confusion and to assign responsibility for the performance of necessary tasks.

8.7.2 Before receiving a bomb threat:

1. Carefully follow opening and closing procedures.
2. Observe and report any unusual or suspicious behavior or incidents.
3. Know what to do during a bomb threat, and your specific assignments during emergency responses.
4. Do not discuss company business with customers or other persons.
5. If the facility is so equipped, use the suspicion camera to record unusual or suspicious persons or events.
6. Insure that the workstation in use contains all emergency forms, and that the surrounding areas are kept free from debris and other items that may conceal a bomb.

8.7.3 During a bomb threat:

1. The employee receiving the threat is responsible for completing the appropriate form, asking all questions if possible. The primary purpose of the person receiving the threat call is to keep the caller on the telephone line while obtaining as much information as possible.
2. Once an employee has established contact with the caller, that employee will remain the sole contact person for any subsequent calls whenever possible. The exception to this policy is when the caller agrees to conduct further conversations with the Unit Security Officer.
3. The line on which the call was received will remain open for the duration of the event and any notifications will be made on alternate telephone lines.
4. When possible, two employees will monitor the threat call, and each employee will be responsible for completing the appropriate form. The employee receiving the call will remain the only person talking with the caller, however.
5. After receiving the initial message, the employee will ask the caller to repeat the message for clarity.

8.7.4 After the threat:

1. Upon termination of the call by the caller, the employee receiving the call will immediately notify the Unit Security Officer.
2. This employee will be assigned to the supervisor handling the threat call for the duration of the incident.

3. All other employees will await further instructions from the Unit Security Officer.
4. If the facility is so equipped, no employee will activate the robbery alarms in response to any bomb threat call.
5. As soon possible, the employee receiving the bomb threat call is to be removed from the work area and isolated. The employee will be asked to provide any and all detailed information to the supervisor to whom he/she is assigned. When this has been completed, the employee is to resume the responsibility of talking with the caller should he/she call again, if appropriate.
6. Any other employees witnessing unusual or suspicious activities within or around the office during the call are to immediately report this information to the Unit Security Officer.
7. The Unit Security Officer will request that all employees visually search the areas immediately surrounding their workstations and report objects that are suspicious. Any employee locating a suspicious object is responsible for reporting that object only: no employee will otherwise disturb the object.
8. If the office is evacuated, all persons leaving the building will be guided to the appropriate emergency staging area(s) by members of the emergency evacuation team. Wherever possible, witnesses are to be identified and asked to remain at the emergency staging area(s), pending contact by the law enforcement agency.
9. All employees will remain at the emergency staging area(s) until they are authorized to leave by the Unit Security Officer or the Security Director.
10. All requests for interviews by the media are to be referred to the law enforcement agency or the Security Director.

8.8 EXTORTION PROCEDURES: MANAGEMENT

8.8.1 PURPOSE: The purpose of this section is to define and describe acceptable management procedures for preventing or responding to an extortion attempt, to avoid confusion and to monitor the performance of necessary tasks assigned to staff personnel. In addition to the instructions detailed for robbery response by staff and management personnel, this information is designed to reduce the possibility of becoming a ransom victim and, if an employee or family member becomes such a victim, to enhance the probability of that person's safe return.

8.8.2 <u>Before</u> receiving an extortion call:

1. Insure that appropriate opening and closing procedures are carefully performed.
2. Act upon reports of unusual behavior or incidents.

3. Conduct frequent extortion response and hostage procedures training during staff meetings and other appropriate occasions.
4. Encourage staff personnel to take appropriate actions regarding suspicious persons or events and to report them promptly.
5. Insure that forms assigning employees to specific responsibilities are current and complete, and that each workstation contains adequate emergency forms. Also verify that each employee knows what is expected from him/her during such events.
6. The Unit Security Officer will create a simple hand signal system that will alert office personnel to an extortion telephone call, or that some other emergency situation is in progress.
7. All employees will be advised at the time of their initial orientation that:
    1. The company will pay ransom demands according to specific policy and procedures.
    2. The company will make every reasonable effort to delay any such transactions.
    3. The company will cooperate with law enforcement agencies at all times.

8.8.3 <u>During</u> the extortion call:

1. If you are notified of the emergency during the extortion telephone call, assign the employee receiving the call to remain in contact with the caller if possible.
2. Immediately notify the Unit Security Officer, local law enforcement agency, the Security Director and the FBI of the call by using another telephone line.
3. Avoid any actions that might jeopardize employees or customers, or that might alert customers or other persons in the office that emergency situation exists.
4. Dedicate specific telephone lines as emergency lines and station an employee with those telephones to answer them if necessary, and to insure that the lines are not used for any other purpose until the incident has been terminated.
5. In the event a hostage is taken inside or brought into the office, the following procedures will be immediately placed into effect:
    1. Insure that all personnel respond in a manner appropriate to maintaining the safety of the hostage and all other persons;
    2. Tell the suspect that you will cooperate with all demands;
    3. Comply with all orders and instructions given by the suspect without delay;
    4. If possible, repeat the instructions back to the suspect before acting, in an effort to make all communication clear;

5. Do not activate alarms: notify a law enforcement agency by telephone only if it is safe to do so;
6. Encourage the suspect to leave the hostage in the office after all demands are met, if possible; and
7. After the robber has left the office, insure that all procedures outlined in the robbery procedures section have been completed.

8.8.4 <u>After</u> the extortion call:

1. Remove the employee receiving the call to a private area within the office, assign another employee to serve as an advocate and to stay with the affected employee, and isolate both employees.
2. Collect any completed witness statements and available evidence and turn them over to the law enforcement agency or the Security Director upon his/her arrival, and insure that all employees have completed their respective tasks as necessary.
3. Insure that all employees remain at the office until authorized to leave by the Unit Security Officer or the Security Director. If a prolonged investigation necessitates employees remaining within the office, those employees will be allowed to contact their families to report their safe condition, if desired. An employee who has been kidnapped or held hostage will have his/her family notified in person by a company executive.
4. The company's Executive Committee, composed at least of the Unit Security Officer, any two (2) Executive Officers and the Security Director will evaluate the threat level and related circumstances, to determine how and if the ransom is to be paid. The Security Director has the final authority: if any of the other three individuals are absent during the threat, then the remaining members will make the decision.
5. The Security Director is responsible for making all arrangements, further contacts when possible and for arranging actual payments to the suspects.
6. Payment of any ransom or extortion demand will only be made or arranged by the Security Director, and in no instance will approval of payment be made by any employee who is personally affected by the event.
7. All media inquiries will be handled by the Security Director.

8.9 EXTORTION PROCEDURES: STAFF

8.9.1 PURPOSE: The purpose of this section is to define and describe acceptable staff procedures for preventing or responding to an extortion attempt, to avoid confusion and to assign responsibility for the performance of necessary tasks. In addition to the instructions detailed for robbery response by both staff and management personnel, this information is designed to reduce the possibility of becoming a ransom victim and, if an employee or family member becomes such a victim, to enhance the probability of that person's safe return.

8.9.2 Before receiving an extortion call:

1. Carefully follow opening and closing procedures.
2. Observe and report any unusual or suspicious behavior or incidents.
3. Know what to do during an extortion attempt and know assignments during emergency responses.
4. Do not discuss company business with customers or other persons.
5. Use the suspicion camera to record unusual or suspicious persons or events.
6. Insure that the workstation in use contains all emergency forms.
7. Refer inquiries by telephone or by person(s) under suspicious circumstances, where specific personnel information is requested, to the Unit Security Officer.
8. Employees are encouraged to vary their daily routines such as the route driven to work, lunch and break times, and other factors that may lead to the establishment of an identifiable pattern.
9. No employee will be permitted to meet customers or other persons at a location that would place that employee in jeopardy.
10. When an employee is assigned to transport documents or other materials from one office to another, or when an employee must, by reason of job description, travel from one office to another, that employee will:
    1. Notify the office before leaving that he/she:
        1. Is en-route to another office or assignment;
        2. Is expecting to travel by a certain route; and
        3. Expects to arrive at a designated time.
    2. Call the office of destination with the same information; and
    3. Upon arrival, call the office of origination.

8.9.3 During an extortion call:

1. The employee who receives the threat is responsible for completing the appropriate report form and for asking all questions, if possible. The primary purpose of the person receiving the threat call is to keep the

caller on the telephone line while obtaining as much information as possible.

2. Any employee receiving notification that an extortion attempt has occurred will immediately notify the Unit Security Officer.
3. The line on which the call was received will remain open for the duration of the event: any notifications will be made on alternate telephone lines.
4. If the notification is received by telephone, the employee will suggest that the Unit Security Officer should negotiate with the caller. If the caller specifically demands that the person receiving the call remains the contact person, that employee will be assigned to that task for the duration of the incident, if practical.
5. When possible, the threat call will be monitored by two employees, and each employee is responsible for completing the appropriate report form. The employee receiving the call will remain the only person talking with the caller, however.
6. After receiving the initial message, the employee will ask the caller to repeat the message for clarity.
7. In the event a hostage is taken inside or brought into the office, the following procedures will be immediately placed into effect:
    1. All personnel will respond in a manner appropriate to maintaining the safety of the hostage and other persons;
    2. Tell the suspect that you will cooperate with all demands;
    3. Comply with all orders and instructions given by the suspect without delay;
    4. If possible, repeat the instructions back to the suspect before acting;
    5. Do not activate alarms or notify a law enforcement agency by telephone;
    6. Encourage the suspect to leave the hostage in the office after all demands are met, if possible; and
    7. After the suspect has left the office, complete all procedures as outlined in robbery procedures section of this manual.

8.9.4 After the extortion call:

1. Upon termination of the call by the caller, the employee receiving the call will immediately notify the Unit Security Officer.
2. This employee will be assigned to the Unit Security Officer handling the threat call for the duration of the incident.
3. All other employees will await further instructions from the Unit Security Officer or the Security Director.

4. No employee is to activate the robbery alarms in response to any extortion call.
5. As soon as possible, the employee receiving the extortion call is to be removed from the area, supplied with another employee to serve as an advocate, and isolated. The employee will be asked to provide any and all detailed information to the Unit Security Officer to whom he/she is assigned. When this has been completed, the employee is to resume the responsibility of talking with the caller should he/she call again, if appropriate.
6. Any other employees witnessing unusual or suspicious activities within or around the office during the call are to report this information to the Unit Security Officer at once.
7. If the extortion threat or ransom demand is received in a written form, exercise care in preserving the document and any other evidence. All evidence will be placed in a paper or plastic envelope immediately, and delivered to the Unit Security Officer or the Security Director.
8. All requests for interviews by the media are to be handled by the Security Director.

8.10 MEDIA RELATIONS PROCEDURES: ALL PERSONNEL

8.10.1 **PURPOSE**: The purpose of this section is to define and describe the company's media relations process. The company recognizes that media representatives have the right to investigate and report a newsworthy event. The company's policy is to cooperate with media representatives engaged in a legitimate investigation, using the guidelines stated in this section,

8.10.2 Only a member of the Crisis Management Team will release information to media representatives, preferably in the form of a press release rather than an on-site interview. All other employees are to refer requests for interviews to the Security Director.

8.10.3 If a press release is issued, the release will contain the following information:

1. An initial damage assessment to facilities only, if appropriate. Statements regarding injuries, fatalities and general geographic conditions will be referred to the responsible emergency services agencies;
2. The name(s) of the person(s) responsible for managing the investigative or security effort;
3. The names of persons authorized to release information to the media;
4. A statement that all facilities are open if a facility has been closed, including:
    1. The estimated date and time when the facility will reopen; and

2. The location designated as a temporary customer-service center, including:
    1. The name of the on-site representative(s);
    2. The physical address of the location;
    3. Any "hotline" number designated for customer inquiries; and
    4. The hours of operation.

8.10.4 If an on-site interview is appropriate and necessary, the interview will focus upon the same guidelines as a press release:

1. Media representatives will be treated courteously and professionally at all times. Complaints concerning a media representative's actions will be referred to the Security Director.
2. If any concern exists regarding the release of certain information, that information will not be released.
3. The names of any injured persons or victims will be divulged only by the law enforcement agency.
4. No person will make comments concerning the alarm system, surveillance cameras, security measures, crisis management procedures or a description of the suspect(s) except to the law enforcement agency.
5. Media representatives will not be permitted to interfere in any way with the activities being conducted by either the company or the law enforcement agency investigating the event.
6. Media representatives will not be permitted to photograph or videotape the interior of any office or other workplace, and any other photographs and videotapes will be taken only from a place of public access.
7. Photographs of employees will not be taken without the employee's permission and the consent of the law enforcement agency.
8. Media representatives will not be furnished with an estimate of any monetary or other loss, and only the law enforcement agency will be authorized to release this information.
9. Media representatives may be permitted to use the telephone in areas other than those restricted work areas affected by the incident.
10. Media representatives will be furnished with the name and telephone number of a contact representative of the company, so that they may obtain further information as it becomes available.

8.11 ROBBERY PROCEDURES: MANAGEMENT

8.11.1 PURPOSE: The purpose of this section is to define and describe acceptable management procedures for preventing or responding to a robbery, to avoid confusion and to monitor the performance of necessary tasks assigned to staff personnel.

8.11.2 Before a robbery:

1. Verify that appropriate opening and closing procedures are performed.
2. Encourage staff members to take appropriate actions regarding suspicious persons or events and to report them promptly.
3. Act upon reports of unusual behavior or incidents.
4. Instruct staff members regarding the appropriate use of security devices.
5. Conduct frequent robbery training during staff briefings and other occasions as appropriate.
6. Conduct frequent spot checks of cash stations to minimize loss exposure.
7. Insure that forms assigning employees to specific responsibilities are current and complete and that each work station contains adequate emergency forms. Also verify that each employee knows what is expected from him/her during such events.
8. Verify that each work station is equipped with all necessary emergency forms and telephone numbers.

8.11.3 During a robbery:

1. Avoid any actions that might jeopardize employees, customers or other persons, including attempting to go to the aid of the employee being robbed.
2. Insure that the suspect's orders are obeyed exactly, if you are able to do so.
3. Activate the alarm and cameras only when it is safe to do so.
4. If the responding law enforcement agency establishes telephone contact with the office, assign an employee to maintain contact with the agency if possible and to update the agency on developments.
5. Remain observant concerning details of the suspect and the employees' actions during the robbery.
6. Take notes if this can be accomplished without being observed by the suspect.

8.11.4 <u>After</u> the robbery:

1. Activate the alarm and cameras again, notify the Unit Security Officer of the robbery if he/she has not become aware of the event, and lock all exit doors.
2. Immediately assemble all employees and instruct them to perform their respective emergency tasks.
3. Verify that all employees, customers and other persons are safe, or determine what medical assistance may be required.
4. Remove all persons from the area affected by the robbery, and assign at least one employee who is not engaged in performing special tasks to remain at that affected area, to avoid evidence contamination.
5. Determine which employee will contact law enforcement officers outside the office when the officers arrive, and relay the employee's name and description to the agency before sending the employee outside the office.
6. Remove the employee experiencing the robbery to a private area within the office, provide him/her with an advocate and isolate these employees.
7. Contact customers and other persons remaining within the office and request that they remain on the premises and available until they are released by the law enforcement agency.
8. Verify that the local law enforcement agency is responding, then have an employee contact the Security Director to respond to the victim office.
9. Insure that all media inquiries are handled by the Security Director.
10. Collect any completed witness statements and other appropriate evidence and turn them over to the Security Director upon his/her arrival. Verify that all employees have completed their respective tasks as necessary.
11. If a prolonged investigation requires that employees remain within the office, those employees will be permitted to contact their families to report their safe condition, if desired. Any employee suffering injury during the robbery will have his/her family notified, in person, by a company executive.

8.12 ROBBERY PROCEDURES: STAFF

8.12.1 **PURPOSE**: The purpose of this section is to define and describe acceptable staff procedures for preventing or responding to a robbery, to avoid confusion and to assign responsibility for the performance of necessary tasks.

8.12.2 **Before a robbery:**

1. Carefully follow opening and closing procedures.
2. Observe and report any unusual incidents and behavior.
3. Know what to do during a robbery:
    1. Know assignments during emergency responses.
    2. Study procedures.
    3. Review suspect description forms.
    4. Know the locations and capabilities of security devices.
4. Minimize cash exposure frequently.
5. Do not discuss business with customers, such as:
    1. Cash and employee authorization limits.
    2. Personal matters.
    3. Details of cash handling procedures.
    4. Physical layout of the office.
    5. Security procedures.
6. Use the suspicion camera to record unusual or suspicious persons or events.

8.12.3 **During a robbery:**

1. Stay calm and obey the suspect's orders exactly. Repeat the orders back to the suspect before acting, if possible.
2. Keep your hands in plain view and avoid rapid movements or actions that may be misinterpreted.
3. Give the suspect no more and no less than he/she demands.
4. Activate the surveillance and alarm cameras only when it is safe to do so.
5. Be observant concerning the physical details of the suspect.
6. Keep any demand notes received and remember which areas of the office the suspect touched.
7. If possible, observe the suspect's direction of travel, the number of accomplices and their descriptions, and any other pertinent information.
8. No employee is to perform any action that will endanger his/her or a customer's safety.

8.12.4 After the robbery:

1. Activate the alarm and cameras again and contact the local law enforcement agency by telephone, giving all necessary information and keeping the telephone lines open to the agency.
2. All employees will now perform their specified emergency tasks.
3. The employee experiencing the robbery is to be removed from the area immediately. This employee will be furnished with another employee to serve as an advocate and isolated, and then requested to complete reports and forms as appropriate.
4. The Security Director is to be contacted and requested to respond to the office to assume security responsibilities.
5. All witnesses to the robbery are to complete their own reports and deliver these to the Security Director.
6. All media requests will be handled by the Security Director.

[This page intentionally left blank]

9.0 SAMPLE FORMS

MEMO

To: Crisis Management Team Coordinators

From:

Date:

Subject: Annual Training Meeting

I conducted a site inspection for the ____

Office/Facility/Department on _______________________.

I reviewed the office/department/facility's violence prevention and response program _____________________.

I conducted a training meeting regarding the company's violence prevention and response program on __________________. A copy of the training roster is attached.

Check one:

_______ The office/facility/department is in compliance with all provisions of its part of the program.

_______ The office/facility/department is not in compliance with all provisions of its part of the program. Please contact me immediately to discuss the indicated items:

_______ POLICY OR PROCEDURE:

____

____

_______ EQUIPMENT OR SUPPLIES:

____

____

________________________________ ________________________________

Name Title

[This page intentionally left blank]

BOMB CALL WARNING FORM
Date: Time: Office:
Person receiving call:
Line receiving call: Location in office:
Instructions:
REMAIN AS CALM AS POSSIBLE. DO NOT UPSET THE CALLER. TELL THE CALLER THAT YOU WILL COOPERATE. NOTIFY ANOTHER EMPLOYEE IF POSSIBLE.
Ask these questions
Who is this?
Why are you calling here?
Is this a prank?
Why are you doing this?
Where is the bomb?
When is the bomb set to go off?
How big is the bomb?
What kind of bomb is it?

What does the bomb look like?
Did you put the bomb together yourself?
How did you get the bomb in here?
Do you know that people may be killed?
What is your name?
Are you part of a group?
Where are you calling from?
What do you want me to do?
Do you really have to do this?
Why?
Is there anything else I need to know?
Please tell me who you are.

Caller's facts:	
Male: Female: Child:	Unknown:
Age range:	Local/long distance call:
Identity if known:	
Voice characteristics	
Tone:	Speech:
Language:	Accent:
Manner:	Other:
Names mentioned:	
Locations mentioned:	
Events mentioned:	
Motive mentioned:	
The exact words used by the caller were	

Background noises	
Machinery:	People:
Airplane/train/bus:	
Street traffic:	
Animals:	Music:
Notifications made in order	
By:	Time:
Law enforcement agency:	
Security Director:	
FBI:	
Company administration:	
This form is to be delivered to the law enforcement agency or Security Director at the conclusion of the call.	

INCIDENT LOG FORM		
Chronological Log Of Events		
Date:	Time started:	Time ended:
Office or department:		
Operator:		
Office or department contact:		
Telephone:	FAX:	
Location:		
Alternate contact:		
Telephone:	FAX:	
Location:		
Recovery site:		
Recovery site:		
Recovery site:		
Notes:		

CHRONOLOGICAL LOG OF EVENTS				
Date:	Time started:		Time ended:	
Location:				
Manager in charge:				
Operator:				
Time	Person	Activity/request	Disposition	Time

EXTORTION TELEPHONE CALL FORM
Date: Time: Office:
Person receiving call:
Line receiving call: Location in office:
Instructions
REMAIN AS CALM AS POSSIBLE. DO NOT UPSET THE CALLER. TELL THE CALLER THAT YOU WILL COOPERATE. NOTIFY ANOTHER EMPLOYEE IF POSSIBLE.
ASK THESE QUESTIONS:
Who is this?
Why are you calling here?
Is this a prank?
How do I know that this is not a prank?
Why are you doing this?
May I talk with...?
Is he/she all right?
What is he/she wearing?
What do you want?
How do you want the money prepared?

Who do you want to deliver the money?
Can we exchange the money for ...?
Where should the money be delivered? When?
Are you part of a group?
Where are you calling from?
What do you want me to do?
How will the delivery person recognize you?
If we can't get the money, how do we let you know?
Is there anything else I need to know?
Please tell me who you are.

Caller's facts	
Male: _______ Female: _____ Child: __________Unknown: __________	
Age range:	Local/long distance call:
Identity if known:	
Voice characteristics	
Tone:	Speech:
Language:	Accent:
Manner:	Other:
Names mentioned:	
Locations mentioned:	
Events mentioned:	
Motive mentioned:	
The exact words used by the caller were	

What do you want?
How do you want the money prepared?
Who do you want to deliver the money?
Can we exchange the money for ...?
Where should the money be delivered? When?
Are you part of a group?
Where are you calling from?
What do you want me to do?
How will the delivery person recognize you?
If we can't get the money, how do we let you know?
Is there anything else I need to know?
Please tell me who you are.

Caller's facts	
Male: _______ Female: _____	Child: __________Unknown: __________
Age range:	Local/long distance call:
Identity if known:	
Voice characteristics	
Tone:	Speech:
Language:	Accent:
Manner:	Other:
Names mentioned:	
Locations mentioned:	
Events mentioned:	
Motive mentioned:	
The exact words used by the caller were	

Background noises	
Machinery:	People:
Airplane/train/bus:	
Street traffic:	
Animals:	Music:
Notifications made in order	
By:	Time:
Law enforcement agency:	
Security Director:	
FBI:	
Company administration:	
This form is to be delivered to the law enforcement agency or Security Director at the conclusion of the call.	

COPING WITH CRIMES OF VIOLENCE TRAINING

Training Roster Date: ____________________

Office/Department/Facility: ______________________________________

Printed Name	Signature

Page 1 of 2

COPING WITH CRIMES OF VIOLENCE TRAINING

Training Roster - Date: ___________________

Office/Department/Facility: ____________________________________

Printed Name	Signature
(page 2 of 2)	

EMPLOYEE PERSONAL PROFILE FORM
Profile Information: The information requested in this questionnaire is voluntary and confidential, and is not to be used for any purpose other than during an actual emergency. The contents of this questionnaire shall be kept in a sealed envelope in a secure area, and shall not be opened unless in the case of an actual emergency. The contents of this questionnaire and a photograph shall be updated annually during your performance evaluation.
Personal Identifying Information
Your name:
Nickname or other names used:
Employment classification:
Employment location:
Permanent residence:
Telephone:
Secondary residence:
Telephone:
Other employment, if applicable:
Date of birth: / / Place of birth:
Name of hospital: Mother's name:
Race: Sex: Complexion:
Height: Weight: Hair color: Eye color:
Scars/marks/tattoos:
Hobbies:
Are your fingerprints and a current photograph on file with this company? Yes ___ No ___

Your Family And Emergency Notification Information	
Marital status: Anniversary date: / /	
Name of spouse/roommate:	Nickname:
Name of child:	Birth date: / /
Name of child:	Birth date: / /
Name of child:	Birth date: / /
PERSONS TO CONTACT IN CASE OF EMERGENCY	
Name:	Telephone:
Address:	Relationship:
Name:	Telephone:
Address:	Relationship:
Name:	Telephone:
Address:	Relationship:
YOUR IMMEDIATE CLOSE RELATIVES	
Name:	Telephone:
Address:	Relationship:
Name:	Telephone:
Address:	Relationship:
Name:	Telephone:
Address:	Relationship:
OTHER PERSONS LIVING OR WORKING IN YOUR HOUSEHOLD	
Name:	Relationship:
Name:	Relationship:
Name:	Relationship:
Name:	Relationship:

Name:		Relationship:
YOUR MOTOR VEHICLES		
Year: Make:	Model:	Color:
License:	Driven by:	
Year: Make:	Model:	Color:
License:	Driven by:	
Year: Make:	Model:	Color:
License:	Driven by:	
Year: Make:	Model:	Color:
License:	Driven by:	
YOUR MEDICAL INFORMATION		
Physician:		
Address:		Telephone:
Physician:		
Address:		Telephone:
Hospital:		
Address:		Telephone:
Blood type:	Allergic to:	
Medical condition(s) requiring treatment or medication:		
Treatment or medication:		
Medical condition(s) requiring treatment or medication:		

Treatment or medication:
I authorize my physician(s) to release confidential information in the event of an emergency situation requiring treatment. Signed: Date:
EMPLOYEE:
After completing this form, place the form into the envelope provided and seal the envelope. Write your name (last, first, middle initial) and Social Security number on the face of the envelope. Deliver the sealed envelope to your reviewer. Please include any other information you feel is necessary on a separate page, to be included with this profile.
REVIEWER:
Ensure that the employee's envelope is securely sealed and completed as indicated. Initial and date the face of the envelope and store in a secure area.

SUSPICIOUS INCIDENT REPORT FORM		
Office		
Date:	Time:	Area:
Employee:		
Employee:		
Description of suspicious activity		
Witnesses		
Name:		
Address:		
City/state/zip:		
Home telephone:		Work telephone:
Name:		
Address:		
City/state/zip:		
Home telephone:		Work telephone:
Unit Security Officer notified		
Date:	Time:	In person or telephone:
Law enforcement agency notified		
Date:	Time:	In person or telephone:
Complete a SUSPECT DESCRIPTION FORM describing each suspicious person, vehicles, etc. and deliver this form to the officer receiving this report. A copy of this form is to be retained in the office, and the original forwarded to the Security Director within 24 hours of the incident.		

[This page intentionally left blank]

SUSPICIOUS OR THREATENING TELEPHONE CALL FORM
Date: Time: Office:
Person receiving call:
Line receiving call: Location in office:
Instructions
REMAIN AS CALM AS POSSIBLE. DO NOT UPSET THE CALLER. TELL THE CALLER THAT YOU WILL COOPERATE. NOTIFY ANOTHER EMPLOYEE IF POSSIBLE.
Ask these questions:
Who is this?
What number were you calling?
Why are you calling here?
Is this a prank?
Why are you doing this?
Do you have a complaint to talk about?
Will you talk with my supervisor?
Will you talk with anyone else?
Do you know that we're being recorded?

What do you want?
Are you part of a group?
Where are you calling from?
What do you want me to do?
If the caller threatens an employee, ask these questions:
What has (person) done to make you angry?
Do you know (person) personally?
How do you know (person)?
Would talking with (person's) supervisor help?
Is there anything else I need to know?
Please tell me who you are.

Caller's facts	
Male: Female: Child: Unknown:	
Age range:	Local/long distance call:
Identity if known:	
Voice characteristics	
Tone:	Speech:
Language:	Accent:
Manner:	Other:
Names mentioned:	
Locations mentioned:	
Events mentioned:	
Motive mentioned:	
The exact words used by the caller were	

Background noises
Machinery: People:
Airplane/train/bus:
Street traffic:
Animals: Music:
Notifications made in order
By: Time:
Law enforcement agency:
Security Director:
FBI:
Company administration:

This form is to be delivered to the law enforcement agency or Security Director at the conclusion of the call.

[This page intentionally left blank]

10.0 TRAINING PROGRAM: COPING WITH CRIMES OF VIOLENCE

1.0 INTRODUCTION

Your company's Security Director is responsible for managing the company's Security Program. This Security Program is designed to create and to maintain a safe and secure environment for the company's employees — like you — to work, and for the company's customers to conduct business. You are just as responsible for the company's Security Program as is the Security Director, and your role is a simple — but critical — one. Your role has three (3) parts:

1. To be aware of the need to protect yourself, your family, other employees, customers and other persons on the premises.
2. To observe and report unusual or suspicious persons and events to your supervisor or to the Security Director.
3. To take appropriate action if you feel that your safety — or the safety of another person — is at risk.

This training program is about options — not absolutes — for handling both personal and business issues. You have to understand the personal issues first before you can apply the personal solutions to business issues. Having options, or a "menu" of available choices — and knowing what those options are and how to use them — is important to your safety. The more options that you have available means that you can be more flexible and versatile when responding to emergency events.

Most people never plan to become the victim of a violent act, and you may live your lifetime without having such an experience. The company does not want you to become a victim — either at work or at home. We do want you, however, to think about what might happen to you if certain events do occur, and then also to plan for what you might do. We have made similar plans as a company and we believe that you should do this as our employee.

The company wants you to make bright, intelligent and informed decisions about your safety. We know that people typically make one of these types of mistakes that cause them to become victims: mistakes that are ignorant, naive or stupid. Consider these definitions:

- **Ignorant:** We have no knowledge (training) about this type of event or the persons involved.
- **Naive:** We have no personal experience with this type of event or with the persons involved.
- **Stupid:** We know the "right" thing to do — and we do something else, instead.

People make these types of mistakes because they don't care about the consequences of their actions; because they have been trained improperly; or because they use faulty reasoning to reach conclusions. The company does not want you to make mistakes that may put you — and others — at risk. We want you to use the guidelines contained in this training program to identify the ways that you might be at risk, and then to develop priorities and a planned approach for resolving these risks. Experts agree that the most effective response priorities are based upon the acronym (P-A-C-E):

- **Prevention**: Prevent the opportunity for an offender to make you a victim;
- **Avoidance:** If your prevention efforts are ineffective, avoid contact with the offender.
- **Confrontation:** If you are unable to avoid the offender, confront him/her at a time and place of your choosing.
- **Escape:** Whether confronting the offender works or not, you must now concentrate on escaping to safety.

This training program teaches you how to make better decisions by:

- Giving you examples of events that may occur that might require your response.
- Describing the techniques and tactics used by offenders.
- Proposing behavioral options for you to consider.
- Encouraging you to share and discuss this information and methodology with your family members and co-workers.

When you have completed this training program, the company wants you to ask yourself, "What have I learned to do if I am confronted by one of these situations — and how will I apply what I've learned?"

2.0 WHAT CRIMES OFTEN LEAD TO VIOLENCE?

2.1 Most violent crimes begin with a physical assault, which is the most feared personal crime. Assaults may also be committed mentally and emotionally. A physical "assault" occurs when another person attempts to make bodily contact with you, without your permission or against your will, while intending to intimidate or to harm you. "Simple" assaults frequently lead to more serious crimes. The crime becomes a "battery" — a more serious offense — if the person actually makes bodily contact with you.

2.2 Statistically, one in every four Americans will become the victim of a violent act during your lifetime, either at work, at home — or any place in between. Most of these violent acts begin as a simple assault.

The likelihood of becoming the victim of an assault is present in common — and often related — crimes involving a(n):

- Burglary (e.g., residential, commercial and vehicle).
- Robbery (e.g., "stand-up" holdup, takeover, ATM and carjacking).
- Extortion (e.g., self, family, employee and hostage).
- Relationship (e.g., spouse, child, neighbor, co-worker and customer).
- Terrorist act (e.g., bombing, arson, vandalism and sabotage).
- Random act (e.g., panhandling, pick pocketing, battery, sexual assault and assassination).

3.0 HOW DO OFFENDERS ACT?

3.1 Offenders display offensive, if not criminal, behavior. Offensive behavior is often the first sign of what the offender's true motives might be. Offensive behavior often leads to a violent act. Offenders are selfish and they do not care about your personal feelings, your property or your safety. Offenders generally target people who appear unaware, unprepared, untrained, distracted and vulnerable. Offenders also target people who make mistakes. Understand that combating crime means taking every available measure against it. Become a "hard target" and the offender will likely leave you alone.

3.2 Offenders rely upon the fact that most people are simply not prepared and have not been trained to cope successfully with a violent act — physically, mentally or emotionally. Most people do not see the need for and have never considered developing a personal response plan for coping with a violent act, either on personal or business levels. The company recognizes that applying effective personal safety techniques works in the business environment, too.

4.0 CAN IT REALLY HAPPEN TO ME?

4.1 It can happen to you. Most people — probably including you — believe that "it" will happen to someone else. Don't believe it. No one is exempt. It happens to people just like you every day.

4.2 We make inappropriate decisions every day, either from ignorance or based upon an inappropriate level of trust. We may often — and unknowingly — create the opportunities for an event to occur that makes us appear to be "attractive" victims. Perhaps you have never done any of these things — but many victims have:

- Opening the front door to your home or hotel room in response to a knock, without first looking outside;
- Walking to your car in a dimly-lit parking lot without paying attention to your surroundings (after driving there with your doors unlocked);
- Opening your wallet to a stranger on the street who asks you for money — or turning your back on the person and walking away;

- Arguing heatedly with a stranger in a public place.
- Giving your address and other personal information to a caller conducting a telephone survey — after failing to identify him/her appropriately.
- Getting into an elevator with a person who is physically or emotionally intimidating.
- Remaining in the presence of people who have been drinking — and arguing — in a bar, or becoming intoxicated yourself in the presence of strangers.

5.0 WHAT IF I DON'T WANT TO BE A VICTIM?

5.1 You don't have to be a victim. The single most important factor in preventing any crime is awareness — of your surroundings, your abilities and your options. Be bright and recognize the relationship between cause and effect as this concept applies to your life. If you live your life "on the edge", it's likely only a matter of time until you get hurt. Begin to reduce the opportunity for most violent incidents to develop by practicing these simple response planning guidelines:

- Understand that you are unable to help anyone else unless you first help yourself. Be selfish — think of your own safety, well-being and actions before considering any another person's.
- Learn about and understand the mechanical operation of crimes and the motivation of the offenders who commit them, by participating in a local crime prevention seminar. Listen to people who have already made serious mistakes and who have survived a violent incident, and don't wait to learn about those mistakes first-hand.
- Consciously practice prevention in your daily activities so that you establish and maintain "safe living habits". Recognize and avoid potentially hazardous locations, events and people.
- Participate in a self-defense training program to develop basic physical skills, mental balance and self-confidence. Your display of a self-confident attitude is often sufficient to repel an offender, but remember to underestimate your own abilities and to overestimate your potential opponent's;
- Keep your deterrence and response plans and actions simple to avoid mistakes and confusion. Plan as if everything will go wrong; that help will be unavailable; and that your survival depends solely upon your own efforts.
- Discuss your plan with people that you live and work around, if your response may affect their plans. Survival often integrates quick communication with teamwork, in addition to your individual effort.

5.2 Continue to reduce the opportunity for most violent incidents to develop by practicing these simple response action guidelines:

- Treat every threat as real and every offender as dangerous — and capable of carrying out the threatened action. Underestimate your abilities and overestimate your adversary's.
- Report suspicious and unusual events, conditions or people to your supervisor or to the police as soon as possible. Don't ever assume that someone will just happen by to help you — take the initiative.
- Assume personal responsibility for your role in the event — don't attempt to involve other persons unless it's absolutely necessary. Your risk increases when you involve or rely upon other people in the event.
- Understand that defenders never win. The only hope you may have is to attack your attacker — and don't give up. Sometimes we win; sometimes we lose; and other times we must be content simply to cut our losses.
- Remember that your primary goal if you are attacked is to escape — and that you always run to somewhere instead of running from someone. If you can, run to an open place where other people are present.
- Lying to save your life, if you have to, may be just the right thing to do. If telling the offender a lie will gain you an "edge", use the lie to your advantage.
- The best time to act may not always be now. Choose the best time, place and circumstances to make your escape — even if you have to wait for the best moment.
- Don't be concerned with the offender's health and welfare if you try to defend yourself. Whether or not the offender is injured during the attack isn't your problem.

6.0 WHY SHOULD I HAVE A PLAN?

6.1 A person without a plan and a ship without a rudder have much in common — they are both at the mercy of events beyond their control. Thinking about what might happen to you in certain situations, and then thinking about how you might respond, is just simply good sense. This process allows you to create a plan of action without actually having to experience the event.

6.2 Many people think of a violent act as a simple assault committed by a stranger. Violent acts actually occur in one of three ways: by design, by mistake and by misunderstanding. The primary causes for acts of violence to occur are:

- As the offender's primary goal — by design — and you are the intended target (e.g., a terrorist act, gang initiation rite or during the permanent resolution of a relationship).

- As an unanticipated result — a mistake or misunderstanding — of another action, and you become an unintended participant or an obstacle (e.g., during a burglary, robbery or an argument).

6.3 Consider what you would do if you found yourself involved in one of the six (6) common crimes listed in the Introduction (burglary, robbery, extortion, relationship, terrorist act or random act) and apply this methodology:

- **WHO** Who is likely to become involved in an event, beside yourself? (e.g., the offender, a family member, co-workers, strangers or the police).
- **WHAT** What is the offender's likely goal (motive)? (e.g., take your money, take your car or commit a sexual assault).
- **WHERE** Where is the offender likely to commit the act? (e.g., a shopping mall parking lot, inside a convenience store or at home).
- **WHEN** When is the offender likely to commit the act? (e.g., while driving to work, during a lunch break or during the nighttime).
- **WHY** Why does the offender want to commit this act? (e.g., for revenge, to get money or to get transportation).
- **HOW** How will the offender likely commit this act? (e.g., threaten you with a firearm, use a car, make a verbal threat only).

7.0 WHAT OPTIONS DO I HAVE?

7.1 Every person — including you — should develop some simple techniques, or options, for coping with a potentially violent incident. Victims who have survived violent events generally report that they survived by making logical decisions, rather than by making decisions based upon the emotional aspects of the event. You make a logical decision by using your reasoning abilities, and you make an emotional decision by using your feelings. These victims who survived learned that planning helps you to control your fear and your other emotions until the event is over. These same victims also generally report having developed a defensive plan of action — a strategy. For our purposes:

- A strategy is an approach or long-range plan (policy) carefully designed to resolve a real or anticipated need.
- A tactic is an effective method or technique (procedure) for implementing the strategy.

7.2 There is no "right" or "wrong" behavior when you are responding to a violent event. There are only "appropriate" or "inappropriate" decisions. Appropriate decisions work. Inappropriate decisions do not work — or they make the situation even more hazardous.

Think about it. Developing a plan, even if you never use it, will help you to avoid surprise and to decide to physically, mentally and emotionally whether to:

- Fight (to defend yourself);
- Flee (to escape); or
- Freeze (to become immobilized, and unable either to fight or flee).

[This page intentionally left blank]

8.0 IN WHAT TYPES OF INCIDENTS AM I MOST LIKELY TO BE HURT?

8.1 Any crime can turn into a violent event. People who steal bicycles, forge checks and embezzle from their employers don't want to get caught any more than a robber does. The types of incidents that most often result in a person becoming the victim of a violent act, however, are a(n):

- **Simple assault**: The act of making physical contact with you without your knowledge, permission and/or consent. The offender's goal is generally to express fear, frustration or anger.
- **Sexual assault**: The act of forcing you to participate in a sex act without your knowledge, permission and/or consent. The offender's goal is generally to dominate, humiliate and degrade the victim.
- **Burglary:** The act of entering a building or a vehicle intending to take property or to commit some other crime. The offender's goal is generally to obtain property — primarily money, documents and other valuables.
- **Robbery:** The act of taking property from your person or your presence, against your will, by using force or fear. The offender's goal is generally to obtain property — primarily money and other valuable personal effects.
- **Extortion:** The act of obtaining your property, with your consent, by using force or fear. The offender's goal is generally to exchange a person, thing or information — primarily for money.
- **Relationship:** The act of terminating a relationship by using force or fear, or using violence as a means of revenge. The offender's goal is generally to use threats or acts of intimidation, vandalism, sabotage, extortion and terrorism to cause the other person in the failed relationship to comply with his/her demands.
- **Terrorist act:** The act of gaining recognition for a personal or a political cause or action, by means of force or fear. The offender's goal is to use threats or acts of intimidation, vandalism, sabotage, extortion and terrorism to focus upon specific demands.
- **Random act**: The act of committing a violent act, seemingly for no apparent reason. The offender's goal is often to commit a shockingly visible and witnessed violent act to enhance his/her peer image, or to gain personal recognition for a personal or political cause or action. This type of violent act is committed for the "thrill", "rush", or "experience", and is often committed as a gang initiation rite.

8.2 Offenders often threaten their victims as a prelude to violence. If the threat causes the victim to comply, the offender generally won't become violent. Notable exceptions to this tactic are those offenders who commit sexual assaults and acts of domestic violence, where the offender's primary motive is to humiliate and degrade the victim. Most threats of physical violence made during an event result in actual injury to a victim if the victim does not comply — immediately. Violence usually occurs when:

- The offender confronts the intended victim;
- The offender has not thoroughly planned the act;
- The victim has not planned for his/her response to the act;
- The offender wants to eliminate witnesses; or
- An unanticipated event occurs that alters either the offender's or victim's planned response.

9.0 WHY ARE CRITICAL DECISION-MAKING SKILLS IMPORTANT?

9.1 Faulty decisions create faulty — and often unfortunate — results. You are using critical decision-making skills if you apply a logical thought process to reach a solution to a problem. You are not using critical decision-making skills if you let your emotions dictate your actions. If you have never thought about how you would react in an emergency, it's likely that you will make inappropriate decisions. Responding to emergency situations requires that you receive training involving critical decision-making and that you plan your actions carefully.

9.2 Making some of these decisions is uncomfortable. That's natural and normal. However, there are two critical — and particularly unpleasant — decisions that every person should make before a potentially violent event occurs. These decisions will ultimately affect the person's strategy and available options for coping with violent acts:

1. Am I willing to sacrifice my life for another person — and if so, under what conditions?
2. Am I capable of injuring or killing a person to defend my life, or the life of another person — and if so, under what conditions?

9.3 You may never have consciously thought about these issues. That's natural and normal, too. But your choices here are simple — either you make the decisions or someone else may make those decisions for you. If you have children, for example, they likely believe that you will give your life in an exchange for theirs. If this belief is absolutely (unconditionally) true — reinforce it. If this belief is conditionally true (that you will do this depending upon the circumstances of the event) — explain to your children just what those conditions may be.

9.4 No rational person would ever sanction the idea that injuring or killing another person is a good thing — under any circumstances. But if you are capable of doing this as your best solution to the event, you have increased your choices and improved your chances for survival. There are several hypothetical problem-solving scenarios included in the last several pages of this training program. Use them to learn how to make these decisions for yourself.

10.0 WHAT ACTIONS SHOULD I PLAN TO TAKE?

10.1 Planning appropriately and practicing prevention behavior is often sufficient to cause an offender to target someone else. While this may be unfortunate for "someone else", it is not your immediate problem. Your immediate problem is to get the offender to leave you alone. This part of the training program is designed to help you act appropriately during hazardous situations if avoidance behavior fails, and to apply the same methodology that we used earlier in the planning section.

10.2 Consider what you would do if you found yourself involved in one of the six (6) common crimes listed in the Introduction (burglary, robbery, extortion, relationship, terrorist act or random act) and apply this methodology:

- **WHO** Who is likely to help me? (e.g., a family member, co-workers, strangers or the police).
- **WHAT** What tools do I have available to help me? (e.g., my voice, a car horn, my keys or spray repellent);
- **WHERE** Where can I go for help? (e.g., the police station, any store or home).
- **WHEN** When is the best time for me to act? (e.g., before the assault begins when the offender approaches me, during the assault or now).
- **WHY** Why do I need to act? (e.g., for self-preservation, to protect family members or to distract others).
- **HOW** How can I best survive this event? (e.g., by cooperating, by using non-lethal force or by using lethal force.)

11.0 SHOULD I ALSO BE CONCERNED ABOUT WORKPLACE VIOLENCE?

11.1 This training program has primarily addressed personal issues so far, because you have to understand the personal issues first before you can apply the personal solutions to business issues. But violent acts don't occur only in our private lives. We read the newspaper and magazines, watch the news and learn daily that a(n):

- Employee arrives for work too early and interrupts a burglary — prompting a hostage event.
- Customer is stabbed during an ATM robbery.
- Manager is kidnapped and held for ransom — and killed after the money's been paid.
- Ex-employee returns to the workplace and shoots his former supervisor and many other people.

- A bomb placed in a parked vehicle explodes during a peak pedestrian traffic time, injuring and killing many people.
- Group of gang members stage a takeover robbery and execute several employees and customers first — then they finish the robbery.

11.2 Incidents of workplace violence have become an ever-present reality — one of every six violent acts occurs in the workplace. Workplace violence has become one of the most significant occupational hazards that employees are likely to encounter. Managing the threat of and responding to potentially violent workplace events is now among the greatest concerns faced by your Security Director and the company's owners, directors, executive officers and managers.

11.3 Surveys conducted by research companies, law enforcement agencies, private foundations and other groups suggest that recent economic, cultural and political changes are primarily responsible for the marked increase in workplace violence activity. Whatever the reasons are for this upsurge will be debated for years — but the threat and the problem are real and you should be concerned. The company has worked to reduce the likelihood that an act of workplace violence will ever occur, and it depends upon your participation to meet that goal.

11.4 Violence is brought to the workplace by two types of offenders — those who have a connection to the business and those offenders who do not. Employees and ex-employees; customers and ex-customers; managers and ex-managers; current and former romantic partners — these people have a connection to the business.

11.5 Robbers, burglars, rapists, gang members, terrorists and other criminals generally have no connection to the business. Understanding the motives of these two groups can help employees recognize the danger signs that may precede violence.

12.0 ISN'T MY PERSONAL LIFE ONLY MY BUSINESS?

12.1 Your personal life is your business — unless you bring threatening issues from home to the workplace. When an employee's personal lifestyle becomes a company security concern, the company generally has the right — and the obligation — to determine if the threat presents a company problem.

12.2 All employees — including you — are responsible for reporting potentially hazardous circumstances to a supervisor. Often employees fail to report these events because they fear embarrassment, retaliation or disciplinary action. The company has the obligation to act upon reports received and at least to determine their credibility. Reporting potentially hazardous circumstances allows individuals to plan for their safety and for the company to implement security measures to protect the reporting employees and others.

12.3 It is also critical that employees report all potentially dangerous incidents that occur in the workplace, as well as those events that occur in their personal lives that they could bring into the company. Every threat should be treated as if it is real until it can be proven groundless. The issues that employees should report include:

- **Personal problems**: Employees can start practicing safe living and working habits by recognizing, and reporting where appropriate, the dangerous areas within their own lifestyles. Employees should advise supervisors (and appropriate co-workers) of potentially dangerous issues in their lives outside work that could potentially bring acts of violence into the workplace.
- **Co-worker problems**: Conflicts may begin at the worksite because of minor disagreements or misunderstandings. Supervisors and co-workers are often able to resolve these conflicts quickly and fairly — but only if they are made aware of the problem. Several employees may be experiencing similar conflicts with the same co-worker. When supervisors see an emerging pattern of conflict involving a particular employee, they should talk to that employee and resolve those issues creating the conflict, or refer that employee for more help. It is best to resolve problems early rather than to let them escalate into a major incident.
- **Customers and other person problems**: Customers and other persons on the premises, such as vendors, often present special problems. Supervisors and co-workers often do not have the same familiarity with this type of person's personality as they do with other employees. Customers and other persons are usually on company premises by invitation and, unless they break the law, their behavior is largely unregulated. However, if they create a disturbance, they can be asked to leave. If they refuse, they can be arrested for various reasons.
- **Stranger problems**: Persons with no obvious affiliation with the company may appear on the premises to gather visual information so that they can plan future crimes. These unaffiliated persons may telephone the company to gather information that they cannot obtain visually.

13.0 WHAT GENERAL SAFETY STRATEGIES SHOULD I CONSIDER?

13.1 Besides recognizing and reporting danger signs, employees can adopt general safety strategies, learn to de-escalate tense situations and be prepared for possible violence. The four most important steps employees can take to protect themselves from acts of violence are observing, reporting, de-escalating and retreating, such as to:

- Observe people's behavior.
- Report anything unusual or suspicious.
- De-escalate tense situations.
- Retreat to safety if de-escalation does not work.

13.2 Identifying your available security options is important to your safety.

Before you are ever confronted with a potentially violent event:

- Enter and leave the premises in groups.
- Talk with the person who is the designated Unit Security Officer or safety coordinator for your worksite. Discuss your safety concerns and the company's emergency response plan, including your role.
- Learn what the company wants you to do in an emergency. Learn how to use alarm systems, panic buttons or other emergency systems that you may have available to you, and have emergency contact information posted by your telephone.
- Remove objects that could be used as weapons against you, from desks and other areas where you interact with customers.
- Develop an emergency signal or code word that you can use with your small work group.
- Identify an escape route and practice moving quickly from your work area in case you have to retreat.
- Acknowledge all persons present on the premises with courtesy and respect and do not give anyone a reason to feel ignored. If you cannot answer a question or resolve a situation, find someone who can answer or resolve it for you.

13.3 Implementing a defensive strategy is also important to your safety.

If you are confronted with a potentially violent event:

- Avoid physical contact with the offender despite harassment or verbal abuse, but remember that you can defend yourself if you are attacked. Self defense is a basic, personal right.
- Leave the area immediately, if it is possible.
- If escape is possible, notify your co-workers, supervisors, and others on the way to safety.
- If leaving the area is not possible, attempt to de-escalate the event. De-escalating tense situations will often prevent violent confrontations.
- If de-escalation does not work, you must be prepared for possible violence. That means retreating, if possible, or knowing what to do to reduce your chances of being hurt. If you do not know these things, you should learn them. Talk to your Security Department or your local law enforcement agency about what you can do to prepare for and respond to potentially violent incidents.
- Report verbal incidents to your supervisor immediately — and if a physical offense occurs, notify any supervisor or a law enforcement agency as quickly as possible.

- Discuss the incident only with appropriate persons.

14.0 DO WORKPLACE VIOLENCE OFFENDERS SHARE A COMMON PROFILE?

14.1 Violent acts are committed in the public environment by people from all walks of life. Some offenders target specific victims — most do not. Victims of most violent acts are generally victims of opportunity. They are simply in the wrong place at the wrong time. Notable exceptions to this fact are those offenders who commit "stalking crimes" — sexual assaults and acts of domestic violence — where the offender's primary motive is to humiliate and degrade the victim.

14.2 However, offenders who commit violent acts in the private environment — the workplace — frequently display characteristic behaviors that suggest that they will be more likely to commit violent acts in the workplace than other people. These behaviors, also known as "crisis signals" or "predictors", are often displayed by the offender both at work and at home simultaneously, in a pattern of escalation that may suggest a potentially violent personality.

14.3 Displaying one or more of these behaviors does not mean that the person will be violent at work — but these well-documented behaviors are frequently displayed by "at risk" people. These "at risk" people may be family members, co-workers, customers or vendors.

The most commonly identified warning behaviors are when the person:

- Is found to have a history of violent behavior — before or after employment with the current employer — either on or off the worksite. Examples are incidents involving spousal abuse, child abuse or cruelty to animals.
- Shows an unusual interest in or an obsession with weapons. This signal is particularly significant if this behavior starts spontaneously or if it is out of character for the employee. Examples are showing an interest in paramilitary training, weapons collections — often including semi-automatic weapons — and the compulsive reading and collecting of gun and "survival" magazines.
- Talks excessively about weapons at work, carries a concealed weapon or displays a weapon to test a person's reactions. An example is observing a weapon in an employee's purse or vehicle glove box.
- Makes verbal — and often profane — threats, either directly or indirectly. An example is predicting that "bad" things are going to happen to a co-worker or to a supervisor.
- Makes spontaneous statements concerning bizarre — or fanatical — religious or political beliefs or "doomsday" predictions. These statements are particularly significant if the employee is adamant that these are the only "true" beliefs. An example is making a statement that a "righteous God will destroy all non-believers."

- Intimidates others or provokes fear in co-workers or supervisors, either physically or verbally. Examples are the placing of harassing telephone calls, bullying or stalking.
- Is obsessively involved with the job, often with no apparent outside interests, few friends, no support system — generally accompanied by relationships that are constantly in turmoil. An example is when the workplace becomes the person's sole source of identity.
- Is a loner with almost no involvement with co-workers, with the possible exception of an uninvited romantic interest in a co-worker. An example is when the romantic interest becomes an obsession and makes the co-worker feel threatened — and the co-worker then reports the attention as a form of sexual or other harassment.
- Acts paranoid, feels pursued and over-reacts to common events. An example is when an employee states that someone is "out to get" him/her and facts do not support this claim.
- Does not accept constructive criticism, the responsibility for results and never forgets an "injustice". An example is insisting upon being "right" — and that you are always "wrong".
- Expresses desperation about current or recent domestic, financial or personal problems. An example is showing an unwillingness to accept that a failed relationship has ended.
- Fascinated with other recent incidents of workplace violence, generally accompanied by voicing approval of the offender's actions. An example is displaying a newspaper clipping of the incident in his/her work area.
- Displays a willingness to compromise his/her personal values, morals, ethics and personal safety, remaining just within acceptable limits. An example is taking unwarranted safety risks and encouraging others to take those risks also.
- Fails to take consistent and appropriate disciplinary action against employees whom he/she supervises, and who threaten to or who commit acts of aggression against other employees. An example is failing to investigate a sexual or other harassment claim properly.
- Becomes inappropriately distressed during workplace events that naturally cause stress. Examples of these events are pending layoffs, terminations, work reductions or labor disputes.
- Works at a location whose geography and events expose employees to the threat of violence from non-employees. An example is having to travel through a neighborhood controlled by a gang, just to arrive at and leave from work.

15.0 HOW DO I DEVELOP MY STRATEGY AND TACTICS?

15.1 Recognize that the first (and often a continual) step in this process of creating a safe and secure environment is to educate yourself — and those people with whom you live and work — about workplace violence using the time-honored standard: who, what, where, when, why and how.

15.2 You have to know your adversary before you can develop a winning strategy. This training program gives you that information. The company — with your help — can then do many things simultaneously to reduce the opportunity for you, other employees and customers to become the potential victims of violent crimes by determining:

- **WHO** Who is most at risk to these crimes? (e.g., staff members, supervisors, managers, executives, directors — and their families, or customers and vendors).
- **WHO** Who is most likely to commit these crimes? (e.g., transients, area residents, professional criminals, gangs, family members, former employees or dissatisfied customers).
- **WHAT** What crimes are most likely to occur? (e.g., create projections based upon the geographic area, industry and company history and significant changes in surrounding social and cultural structures).
- **WHERE** Where are the most likely sites for specific crimes to occur? (e.g., robbers often prefer to commit their crimes at ATMs and within convenience stores in metropolitan areas).
- **WHEN** When are specific crimes likely to occur? (e.g., most crimes affecting cash-handling functions of retail companies are committed during peak customer traffic times, when distractions may create confusion among customers and a lack of focus among employees).
- **WHY** Why are specific crimes committed? (e.g., burglars burglarize to steal property, robbers rob to steal money, terrorists terrorize to draw attention to a personal or political cause).
- **HOW** How are specific crimes committed? (e.g., robbers use verbal commands and gestures, notes, firearms, bombs and other weapons).

16.0 WHAT SHOULD I DO NEXT?

16.1 Creating appropriate prevention techniques requires that the company acknowledges the potential for workplace violence; prepares strategies to address issues likely to occur; and trains its employees to behave appropriately during an event if your prevention efforts are unsuccessful. Develop your prevention strategy using the information you gathered by following a few simple guidelines.

16.2 You can do many things to reduce the opportunity for an act of workplace violence to occur, among them:

- **INVESTIGATE** Learn of the problems experienced by other companies — particularly similar companies — and the techniques they have used successfully to reduce risks, by starting or participating in a local crime prevention group.
- **IDENTIFY** Examine each worksite — not just cash-handling facilities — and the areas surrounding those worksites for attractive hazards that may actually encourage crime (e.g., poor lighting, inadequate visibility or isolated locations).
- **ASSESS** Review your operating policies and procedures and training programs for misleading or incorrect information that employees may have to rely upon during an emergency.
- **COMMUNICATE** Create a communication policy and follow up procedure by which employees may report potential safety and other risks, and that also requires that they report to your Security Director acts involving threats or intimidation that originate either at work or at home. Insure that this process includes a mechanism for monitoring, compliance and correction of the problem.
- **TRAIN** Invite local law enforcement and other agencies to present training programs about domestic violence, sexual assault and other forms of personal assault prevention techniques. This additional training program may help employees to recognize the symptoms displayed by offenders, the likely events that may occur and appropriate prevention and resolution techniques.

17.0 CONCLUSION

Everything you need to make intelligent and informed decisions about your personal safety is not contained here. You can find other useful strategies and tactics by reading the works of many authors; by attending training seminars; by consulting with your local law enforcement agencies; and by discussing these issues with your family and co-workers. Whenever possible, learn from other people's experiences — so that you do not have to learn a lesson yourself.

18.0 HYPOTHETICAL PROBLEMS: COULD THESE HAPPEN TO YOU?

18.1 Understand that you can do everything "right" and still lose and everything "wrong" and still win — depending upon the time, place and circumstances surrounding an event. You may not be able to control the event but you can control your response to it, based upon your education, training, experience — and your appropriate use of this information.

18.2 Test your responses to the following problems taken from real-life examples either individually or as part of a group. There are no "right" or "wrong" solutions — just some solutions will work better than others. Use the suggested decision-making process to determine the most appropriate course of action for yourself, and for the other participants mentioned. Place yourself in each role within each problem. Remember — this is only a test. Use a logical process, not an emotional one. What would you do and how would you do it?

18.3 Personal Issues

You have total responsibility for results in each of the following personal situations. Coping with potentially life-threatening events, either actively or passively, requires courage. Courage is simply doing whatever needs to be done. An act of bravery is simply displaying courage — at great personal risk. The purpose of these exercises is to:

- Cause you to become and remain aware of your surroundings.
- Cause you to recognize that there may be several options available to you, if you think the problem through in a reasonable and logical manner.
- Cause you to recognize that your decisions will affect the decisions of the other persons involved, and to encourage you to discuss your decisions and your plans with those people who may be affected.
- Cause you to make difficult decisions about your personal role or course of action, where any decision may create a loss of some kind.

Suggested answers are available in the last few pages of this training program. As you examine each of these situations, it's important to emphasize a very basic, but often forgotten, reality.

Sometimes we win; sometimes we lose; and sometimes we must be satisfied just to cut our losses. Analyze and respond to these real-life scenarios:

1. **Logic**

   As you drive home from work, you witness a single-car rollover accident involving a two-door car on your side of the highway. This highway is seldom traveled; there are no other motorists in the area; and there are no telephones nearby. You stop to help. You approach the driver's side of the car and observe that the car contains four unconscious people whom you do not know. These people are unconscious; they all have suffered identical, life-threatening injuries; and they cannot help themselves. You can — perhaps — save one life, but the other people will likely die without your help.

   Which one of these people in the two-door car will you choose to help first — and why:

   1. The 28-year-old woman in the driver's seat;
   2. The 70-year-old male minister in the right-front passenger seat;
   3. The 60-year-old woman in the left-rear passenger seat; or
   4. The 18-month-old infant strapped into a car-seat in the right-rear passenger seat?

2. **Burglary**

   You arrive home after work and, as you drive into the driveway, you observe that the front door to your house is wide open. You know that neither your spouse nor your children should be home. You don't see any car parked in your driveway or any people in the area. What will you do? What do you want your spouse and your children to do if they came home first?

3. **Robbery**

   You and your spouse walk from a public parking lot to a movie theater nearby. As you walk around the corner of the theater building, a man steps out into your path. He points a gun at you and says, "This is a robbery. Give me all of your money and you won't get hurt." You don't see anyone else in the area. What will you do? What do you want your spouse to do?

4. **Extortion**

   Your 9-year-old son walks to elementary school every school-day morning at 7:30 A.M. On a Wednesday morning at 7:45 A.M., you receive this call: "We kidnapped your son while he was on his way to school. Get $25,000 in cash, no bills larger than $50s, put the money in a brown paper grocery sack and put the sack next to the Federal Express pickup box at the northwest corner of Alameda and Terry.

Don't call the police, don't let anyone know what you're doing, and don't be followed. Deliver the money, and we'll let your son go in an hour. Don't get the money, and you'll never see him again." What will you do? What do you want your spouse and children to do if you were the victim?

5. **Domestic Violence**

   You are visiting with your best married friends at their home. During your visit your friends begin a violently verbal and physical dispute. They strike each other, causing some physical injury and escalating the emotional impact of the event. What will you do? What do you want your spouse and children to do if they were with you? What do you want your children to do if they were visiting a friend's home without you?

6. **Terrorist Act**

   You begin receiving cards, letters, telephone calls and presents from a "secret admirer". You observe that this person drives by your home several times a day; dines at the same restaurants; shops at the same stores; attends the same events; and generally appears at the same locations — and at the same times — as you do. You recognize that you are being stalked. What will you do? What do you want your spouse and children to do?

7. **Random Act**

   You enter your local video store to return a rental video tape. There are several other persons in the store. As you walk through the store to make another selection, three youthful gang members enter the store carrying guns. As one gang member robs the store, the other gang members begin randomly shooting employees and shoppers. There is no one to call for help. What will you do? What do you want your spouse and children to do? What do you expect the other people in the store to do?

8. **Sexual Assault**

   While walking alone to your car in the shopping mall parking lot at night, you witness the beginning of a sexual assault. Two men are assaulting one female victim. You are the only witness and you do not observe anyone else in the immediate area. What will you do? What do you want your spouse and children to do if they were with you? What do you want your children to do if they were alone?

18.4 Business Issues:

You have total responsibility for results in each of the following business situations. The suggestions for making decisions regarding personal issues also apply to these business issues.

Analyze and respond to these real-life scenarios and prioritize your response according to:

- People (employees and customers).
- Places (facilities).
- Things (assets and records).

1. **Burglary**

   You arrive for work on a Monday morning at 7:30 A.M. with other employees, and you enter the building. You immediately observe that the office has been burglarized over the weekend. Burglars have apparently removed cash, negotiable documents and records. What will you do? What do you expect the other employees would do?

2. **Robbery**

   On a Tuesday afternoon at 12:35 P.M., a nicely-dressed woman walks up to you and hands you a note. The note reads, "This is a robbery. Give me all of your money. Don't push the alarm, and don't yell for help. I won't hurt you if you do what you are told. I will hurt you if you don't follow instructions." What will you do? What do you expect the other employees will do?

3. **Kidnapping/Hostage**

   You arrive for work on a Friday morning at 8:00 A.M., just minutes after the opening employees are scheduled to inspect the building. You observe that the required "all clear" signal is not displayed. You look into the office from outside. You see that a gunman has taken your two best friends hostage, and is yelling to you, "Get inside, or I'll kill 'em." What will you do? What do you expect the hostage employees will do? What do you expect other employees will do if you were the victim?

4. **Workplace Violence**

   On a Monday morning at 10:30 A.M., a man walks into the office and up to your desk. You recognize the man as the recently-divorced husband of one of your fellow employees. You are also aware that the husband has behaved violently and has delivered death threats to this employee in the past. The man is wearing a hunting vest, with what appears to be several sticks of dynamite and wires protruding from the cartridge loops and pockets. The man says to you, "Bring my wife out here. I want to talk to her. Do it now and you won't get hurt, and neither will she." What will you do? What do you expect the other employees will do?

5. **Bomb Threat**

   On a Thursday afternoon at 5:15 P.M., your company is holding an open house in the new office building's lobby. All of your employees are present, as well as several company executives and prominent members of the community. The receptionist answers the telephone as it rings, and hears, "Great party you're having. Because you didn't invite me, I planted a bomb in your new office. The bomb will go off at 5:30, and I won't tell you where it is." The caller then disconnects. What will you do? What do you expect the receptionist to do? What do you want other employees to do?

18.5 Suggested Answers

There are no "right" answers to these questions. There are only those answers that are appropriate (they work) for you, or answers that are inappropriate (they don't work for you or they may make the situation even more hazardous). The best probable answers are described below and these answers include the logical reasons for making these decisions. You don't have to agree — you don't even have to like them. You should always choose the answer that works best for you, because you're the person who has to live with the results.

Personal Issues

1. The most logical answer is the driver of the two-door car. You are already standing at her door and you must remove her to get to the other people. You may be able to revive her and gain an ally so that you can go back and help another person in the car. The people to help — in order of priority — are numbered 1 - 4 in the original problem.
2. You should not go into the house. You should drive to a neighbor's house or to another telephone nearby — and call the police. Your spouse and your children should do the same thing. Burglary is a property crime — but a frightened burglar can turn the event into a violent, personal crime if you surprise him/her.
3. Both you and your spouse should hand your wallet or purse to the robber, without hesitation. Don't try to negotiate for your credit cards, your favorite photographs or other belongings that may also be inside your wallet or purse. Robbery is primarily a property crime and robbers generally don't want to hurt their victims — but they will hurt you if you don't comply with their demands.
4. You should agree to do whatever the kidnaper/extortionist tells you to do. Write notes if you can and ask the caller to repeat the instructions so that you don't make any mistakes. Ask to talk with your son if you can. After you and the caller are disconnected, go to another telephone and call the police. Tell them everything that you can remember about the call and cooperate with the agency's requests. You can't help your son by doing exactly as the caller tells you to do. If you go, no one else knows about the event and you are now at risk, too. And if you're the victim, your family should also follow this plan. Extortion is both a financial crime and a personal crime. Greed and the fear of identification may drive the extortionist to hurt you.

5. You should leave the house immediately after the argument becomes physical and call the police from a telephone nearby. Domestic combatants often turn their anger to bystanders after they are finished with their primary targets. Your spouse and your children should do the same thing. Your children should grab your friends' children, if possible, and leave the house with them. Children are often the most attractive targets of domestic violence.
6. You should report your observations immediately to your spouse, to your employer and to the police. You should also discuss your observations with your children. Neither you nor your spouse should confront the stalker with your observations, or threaten him/her in any way. Stalking is a police issue unless you're forced to defend yourself before the police arrive, and stalking is often a prelude to a sexual assault or to a kidnap/extortion event.
7. You should immediately find a place to hide that is not obvious or easily-discovered. You should not do anything to draw attention to yourself. Your spouse and your children should do the same thing. Trying to negotiate with the offenders is generally fruitless because they enjoy hurting you. Causing the violence is the offenders' primary motive — and the robbery is a secondary one. The other people in the store will likely panic and make themselves ready targets. This distraction will often shift the focus from you to easier victims.
8. You should go inside the nearest mall business and tell someone to call the police. Wait inside the business until the officers arrive. If this isn't possible, you should go to your car, lock the doors and start it. Then turn on the headlights and begin honking the horn to attract attention. If the attackers leave their victim and begin to approach you, leave the area and call the police from the nearest telephone. Your spouse should do the same thing. Your children should simply go to the nearest telephone and call the police.

Business Issues

1. You and the other employees should leave the building immediately and call the police. You should not go back into the building until the police have searched it. You should use opening and closing procedures requiring that one employee searches the building before other employees enter — and after other employees leave — while another employee waits outside the building to call the police, if necessary.
2. You should hand the money to the robber, without hesitation. Ask the robber questions if you don't understand what he/she wants, and repeat demands back to the robber to avoid miscommunication, if you can. You should not activate the alarm until you're convinced that doing so is safe.

Other employees should help you by leaving you alone and by complying with your directions. Remember — robbery is primarily a property crime and robbers generally don't want to hurt their victims. But they will hurt you if you don't comply with their demands.

3. You should not go into the building and you should immediately get out of the robber's view, instead. If you go into the building, the next person to arrive will not know of the problem. You can help your friends by staying nearby; by directing arriving employees to keep the building's door under observation and to keep other people from entering the building; and by directing other people to call the police. Remember — extortion is both a financial crime and a personal crime and allowing other people to enter the building will likely confuse and frustrate the kidnaper.
4. You should agree to get his former wife and to return with her. You should then leave the area and take other people — including his former wife — with you, if you can. You should tell everyone to leave the building and to telephone the police from the nearest telephone. You should not bring his former wife back to him. This "domestic terrorist" has one objective — to bring attention to his failed marriage by holding everyone in the building hostage. He will also likely commit suicide and kill his former wife, as well.
5. You should immediately tell one of the executives of the threat. The executive should order an immediate evacuation of the building, sending all people to the building's designated emergency staging areas. The executive should then send several employees to call the police from the nearest telephones located outside the building. Employees should visually — not physically — search the building before they leave and report their observations to the officers when they arrive.

[This page intentionally left blank]

19.0 LEADER'S GUIDE: COPING WITH CRIMES OF VIOLENCE

LEADER: Department or Office Leader

DISCUSSION TIME: 45 minutes

19.1 OVERVIEW

The company has developed and implemented a comprehensive violence prevention and response program for all personnel, departments, offices, facilities and functions. Training the company's personnel appropriately to implement the program is critical to its success.

19.2 OBJECTIVES

- To inform all personnel about the company's violence prevention and response program, and the policies and procedures regulating its implementation on a company-wide scale.
- To advise all personnel about the company's violence prevention and response program, and the policies and procedures regulating its implementation, as the program affects specific functions, offices and departments.
- To train each employee to prevent violence or to respond appropriately during a violent incident, according to specific instructions designed for the employee's job classification.
- To suggest safety precautions for use in both a business and personal environment.

19.3 DISCUSSION MATERIALS

- Handout: Coping With Crimes of Violence Training Handout.
- Emergency staging area and facility diagram.
- Violence policy.
- Media policy.
- Prevention and response procedures (position-specific).

19.4 MEETING PREPARATION

- Conduct an interior and exterior inspection of the facility, noting potential hazards for discussion.
- Note locations of electrical and water connections, if appropriate. Do not attempt to test the switches or valves. Determine if any shut-off device requires a tool to operate: if so, determine if a tool is readily available.
- Verify compliance with local codes regarding electrical outlets, exit sign devices, hazardous materials storage standards, and other appropriate requirements (a local fire marshal has a variety of inspection checklists).

- Verify that the emergency staging areas are correctly designated on the plan diagram, and that the areas selected are free from inherent hazards (i.e., overhead electrical wires, fire plug, gas main, etc.).
- Verify that lists containing information regarding the facility's personnel, emergency services and related agencies, equipment and appropriate records are current and available.
- Verify that first aid kits and fire extinguishers are available and functional.
- Verify that an appropriate inventory is maintained of all necessary forms and supplies.
- Designate appropriate employees to perform necessary functions during a violent incident:
  - Evacuation of other employees and customers.
  - Securing assets and records.
  - Securing the facility and utilities.
  - Record-keeping and report writing.
- Prepare for the meeting by:
  - Reading the handout to familiarize yourself with the information, and to insure that you understand the key elements of the violence prevention and response program. Discuss questions or concerns with your supervisor.
  - Distributing the handout and any related materials to each employee in your office or department one (1) week before your meeting. Instruct employees to read the handout, note questions or comments directly on the copy, and to bring the handout to the meeting.
  - Assembling all appropriate materials, displays and the handout to use as references during the meeting (have extra copies of the handout available at the meeting).
  - Preparing a meeting outline on a flip chart to help guide your presentation, if appropriate.
  - Preparing a training roster (a copy is attached).

19.5 CONDUCT THE MEETING

- Assign an employee to complete the roster and take notes during the meeting, and to deliver the signed roster and notes to you at the end of the meeting for your report (a device is attached to this guide).
- Introduce the purpose of the meeting, and the training program objectives (refer to your flip chart, if appropriate, throughout the presentation).
- Refer to the handout.

- Define a violent act and discuss the types of potentially violent incidents; the potential impact upon the company and the employee personally. Emphasize the importance of preparation.
- Discuss the primary goals of violence prevention and response and how important the program is to the company and its employees.
- Discuss the program and emphasize the role and responsibilities of each employee.
- Discuss the facility's emergency staging areas and evacuation routes, and designate the employees responsible for evacuation.
- Discuss the facility's utilities (water, electricity and gas) and their appropriate locations; emergency telephone numbers and their appropriate use; and the potential re-location of staff to an alternate work site, and describe those designated sites.
- Discuss public and media relations, and each employee's role in protecting the company's image.
- Discuss personal responsibilities and the potential impact upon business recovery, and the suggestions for personal preparation for a violent incident.
- Ask that any employee having appropriate comments, concerns or questions contact you after the meeting, if an employee is not comfortable discussing these issues during the meeting.
- Distribute the training roster for employees' names and signatures.

19.6 AFTER THE MEETING

- Retrieve the roster and notes from the note-taking employee.
- Using the notes and your personal observation, indicate in writing what issues need to be resolved and your suggestions for resolution.
- Complete the information requested on the attached memo, and return the memo, training roster, and your written report to the Crisis Management Team coordinators.

[This page intentionally left blank]

20.0 SAMPLE MEMOS, FORMS AND FACILITY DIAGRAMS

MEMO

To: Crisis Management Team Coordinators

From:

Date:

Subject: Annual Training Meeting

I conducted a site inspection for the ______________________ Office/Facility/Department on ______________________.

I reviewed the office/department/facility's violence prevention and response program ______________________.

I conducted a training meeting regarding the company's violence prevention and response program on __________________. A copy of the training roster is attached.

Check one:

_______ The office/facility/department is in compliance with all provisions of its part of the program.

_______ The office/facility/department is not in compliance with all provisions of its part of the program. Please contact me immediately to discuss the indicated items:

_______ POLICY OR PROCEDURE:

____

____

_______ EQUIPMENT OR SUPPLIES:

____

____

______________________________ ______________________________

Name Title

[This page intentionally left blank]

COPING WITH CRIMES OF VIOLENCE TRAINING - PAGE 1

Training Roster Date: ___________________

Office/Department/Facility: ____________________________________

Printed Name	Signature

COPING WITH CRIMES OF VIOLENCE TRAINING - PAGE 2

Training Roster Date: ___________________

Office/Department/Facility: ____________________________________

Printed Name	Signature

Page 2 of 2

SEMINAR EVALUATION: COPING WITH CRIMES OF VIOLENCE TRAINING

Seminar name: ______________________________

Seminar location:	
Your name (optional):	
Your title:	
SEMINAR CONTENT & ORGANIZATION (1/POOR 5/EXCELLENT)	**Rating**
1. Will you be able to apply this information in your position?	
2. Was the seminar content well-organized?	
3. Were the handout materials useful and informative?	
4. Were the seminar topics adequately addressed?	
SEMINAR INSTRUCTOR	**Rating**
5. Were questions answered to your satisfaction?	
6. Was the instructor organized and well-prepared?	
7. Was the instructor knowledgeable and up-to-date in the subject matter?	
8. Did the instructor display enthusiasm for the topics and the class responses?	
9. Were you encouraged to ask questions and express opinions?	
10. Were the teaching aids, handout materials and techniques effective?	
Comments	

[This page intentionally left blank]

DISASTER RECOVERY

Contact Directory
500

Emergency Operations Centers

Organization	Phone Number
Alaska Volcanic Observatory	(907) 786-7497
Canada Government Emergency Operations Coordination Center	(613) 991-7000
Department of State Operations Center	(202) 647-1512
FBI Operations Center, Washington, DC	(202) 324-6700
FEMA Disaster Finance Center, Berryville, VA	(540) 542-7300
FEMA Disaster Information Systems Clearing House, Berryville, VA	(540) 542-2189
FEMA Operations Center (FOC)	(540) 665-6100 (703) 771-6100 (800) 634-7084
FEMA Hyattsville National Processing Service Center, Hyattsville, MD	(301) 209-4000
FEMA MERS Operations Center (MOC), Bothell, WA	(425) 487-4449 (800) 395-6042
FEMA MERS Operations Center (MOC), Denton, TX	(940) 898-5280 (800) 260-5110
FEMA MERS Operations Center (MOC), Denver, CO	(303) 235-4847 (800) 311-7021
FEMA MERS Operations Center (MOC), Maynard, MA	(978) 461-5501 (800) 213-8965
FEMA MERS Operations Center (MOC), Thomasville, GA [FEMA Alternate Operations Center (FAOC)]	(912) 225-4756 (800) 792-6196
FEMA Mount Weather Operator	(202) 566-1600
FEMA National Emergency Training Center	(301) 447-1000
FEMA National Interagency Emergency Operations Center (NIEOC)	(202) 646-2470
FEMA National Processing Service Center, Denton, TX	(940) 891-8500
FEMA National Teleregistration Center, Denton, TX	(940) 591-7100
FEMA National Teleregistration Center, Mt. Weather, Berryville, VA	(540) 542-7109
FEMA Virginia National Processing Service Center, Mt Weather, Berryville, VA	(540) 542-7717
Joint Nuclear Accident Coordinating Center (JNACC) Department of Defense (DTRA) Department of Energy	(703) 325-2102 (703) 325-2102 (505) 845-4667

Organization	Phone Number
National Centers for Environmental Predictions Aviation Weather Center Kansas City, MO Storm Prediction Center Norman, OK	 (816) 584-7200 (405) 579-0707
National Coordinating Center for Telecommunications (NCS)	(703) 607-4900
National Hurricane Center	(305) 229-4470
National Interagency Coordination Center Boise, ID	(208) 387-5400
National Interagency Fire Center Boise, ID	(208) 387-5512
National Interagency Fire Desk Washington, DC	(202) 205-1450
National Response Center	(800) 424-8802
National Weather Service General / Severe Weather Info. Local & Extended Forecast (Recording) Riverstage Forecasts (Recording) Washington Weather Info.	 (703) 260-0209 (703) 260-0307 (703) 260-0305 (703) 260-0105
Nuclear Regulatory Commission Operations Center	(301) 816-5100
Office of the Secretary of Defense Crisis Coordination Center	(703) 769-9320
Tsunami Warning Center (Alaska)	(907) 745-5235
Tsunami Warning Center (Hawaii)	(808) 689-6655
Typhoon Warning Center (Guam)	(671) 344-4224
U.S. Capitol Police, Washington, DC	(202) 224-0908
U.S. Marshals Service, Washington, DC	(202) 307-9100 (800) 336-0102
USGS National Earthquake Information Center 24-Hour Recorded Message	(303) 273-8500 (303) 273-8516
White House Situation Room	(202) 456-9431

Military Command Centers

Military Command Center/Organization	Phone Number
Director of Military Support (DOMS) (After duty hours use Army Ops Ctr)	(703) 697-3203
First U.S. Army (Fort Gillem, Georgia)	(404) 469-3280 (404) 469-3288
Fifth U.S. Army (Fort Sam Houston, Texas)	(210) 221-2955
National Airborne Operations Center (NAOC)	(402) 422-1891
National Military Command Center (NMCC)	(703) 697-6340
U.S. Air Force Operations Center	(703) 695-7220
U.S. Army Forces Command (FORSCOM)	(404) 464-5222
U.S. Army National Guard Command Center	(703) 607-9350
U.S. Army Operations Center	(703) 697-0218
U.S. Atlantic Command (USACOM)	(757) 322-6000 DSN 262-6000
U.S. Coast Guard Command Post	(202 267-2100
U.S. Marine Corps Operations Center	(703) 695-7366
U.S. Navy Operations Center	(703) 695-0231
U.S. Pacific Command (USPACOM)	(808) 477-7227
U.S. Transportation Command (TRANSCOM) Mobility Control Center (MCC)	(618) 256-8105

FEMA Regional Offices

Region I

Federal Emergency Management Agency
99 High Street
6th Floor
Boston, MA 02110
(617)-956-7506

Serving:

- Connecticut
- Maine
- Massachusetts
- New Hampshire
- Rhode Island
- Vermont

Region II

Federal Emergency Management Agency
Suite 1311
26 Federal Plaza
New York, NY 10278-0002
(212) 680-3600

Serving:

- New Jersey
- New York
- Puerto Rico
- Virgin Islands

FEMA Regional Offices

Region V

Federal Emergency Management Agency
536 South Clark St., 6th Floor
Chicago, IL 60605
312-408-5500

Serving:

- Illinois
- Indiana
- Michigan
- Minnesota
- Ohio
- Wisconsin

Region VI

Federal Emergency Management Agency
FRC 800 North Loop 288
Denton, TX 76209-3698
(940) 898-5399

Serving:

- Arkansas
- Louisiana
- New Mexico
- Oklahoma
- Texas

Region VII

Federal Emergency Management Agency
9221 Ward Parkway, Suite 300
Kansas City, MO. 64114-3372
(816) 283-7063

Serving:

- Iowa
- Kansas
- Missouri
- Nebraska

FEMA Regional Offices

Region VIII

Federal Emergency Management Agency
Denver Federal Center
Building 710, Box 25267
Denver, CO 80255-02674404
Telephone: (303) 235-4800

Serving

- Colorado
- Montana
- North Dakota
- South Dakota
- Utah
- Wyoming

Region IX

Federal Emergency Management Agency
1111 Broadway, Suite 1200
Oakland, CA 94607-4052
(510) 627-7100

Serving:

- Arizona
- California
- Hawaii
- Nevada
- American Samoa
- Guam
- Commonwealth of the Northern Mariana Islands
- Republic of the Marshall Islands
- Federated States of Micronesia

FEMA Regional Offices

Region X

Federal Emergency Management Agency
Federal Regional Center
130 228th Street, Southwest
Bothell, WA 98021-8627
(425) 487-4600

Serving

- Alaska
- Idaho
- Oregon
- Washington

Federal Headquarters

Federal Emergency Management Agency
500 C Street S.W.
Washington, D.C. 20472

1 (800) 621-FEMA (3362)

TDD: TTY users can dial 1 (800) 462-7585 to use the Federal Relay Service.

Federal Departments and Agencies
24 Hour Contact

Department/Agency	24-Hour Phone	24-Hour Fax
Agency for International Development (AID)	Via DOS Opns Center (202) 647-1512	(202) 647-1811
American Red Cross (ARC)	(703) 206-8823	(703) 206-6296
Canada Government Emergency Operations Coordination Center	(613) 991-7000	(613) 996-0995
Centers for Disease Control and Prevention (CDC)	(770) 488-7100 (800) 311-3435	(770) 488-7107
Chemical Safety & Hazard Investigations Board (CSHIB)	Via FEMA Operations Center (540) 665-6100 (800) 634-7084	(540) 665-6175 (540) 665-6147
Department of Agriculture (USDA) Food and Nutrition Service (FNS)	G. Sheffy Cell: (703) 505-0628 Home: (301) 894-0193	(703) 305-2908
Department of Commerce (DOC)	(202) 482-4128 (4 AM - 8 PM Eastern) **Guards answer after duty hours**	(202) 482-5320
Department of Defense (DOD)	Via Army Operations Center (703) 697-0218	(703) 693-6290
Department of Energy (DOE)	(202) 586-8100	(202) 586-8485
Department of Housing and Urban Development (HUD)	Duty Officer: Bruce McCarthy Home (703) 370-7281	(202) 708-0299
Department of the Interior (DOI)	(P) (800) 345-8859 (A) (540) 999-3422 (2ND Alt) 888-581-2610	(202) 208-5078
Department of Justice (DOJ)	(202) 514-5000	(202) 514-5778
Department of State (DOS)	(202) 647-1512	(202) 647-1811
Department of Transportation (DOT)	Via USCG Cmd Center (202) 366-1863	(202) 366-3768
Department of the Treasury (Treas)	(202) 622-1825	(202) 622-1829
Department of Veteran Affairs (VA)	(304) 264-4800 - Duty Officer: Howard Boyd - (800) 351-2690	(304) 264-4810
Environmental Protection Agency (EPA)	(202) 564-3850 **Via National Response Center** **(800) 424-8802**	(202) 564-8729 Nat. Resp. Ctr. (202) 267-2165
Federal Aviation Administration (FAA)	(202) 267-3333	(202) 267-5289

Department/Agency	24-Hour Phone	24-Hour Fax
Federal Bureau of Investigation (FBI)	(202) 323-3300	(202) 323-2079
Federal Communications Commission (FCC)	(202) 632-6975	(202) 418-2813
Federal Emergency Management Agency (FEMA)	(540) 665-6100 (800) 634-7084	(540) 665-6175 (540) 665-6147
General Services Administration (GSA)	(202) 708-1111	(202) 708-7450
Immigration and Naturalization Service (INS)	(202) 616-5000	(202) 305-4823
National Aeronautics and Space Administration (NASA)	(202) 358-1616 Ask Guard to contact: James Lloyd	(202) 358-3382
National Communications System, National Coordinating Center for Telecommunications (NCS)	(703) 607-4950 Duty Officer, DISA Global Command Center	(703) 607-4998
National Interagency Fire Center	(208) 387-5400 (208) 387-5660	(208) 387-5663
Nuclear Regulatory Commission (NRC)	(301) 816-5100	(301) 816-5151
National Response Center	(800) 424-8802 (202) 267-2675	(202) 267-2165
Office of Foreign Disaster Assistance (OFDA)	Via DOS Opns Center (202) 647-1512	(202) 647-1811
Office of Personnel Management (OPM)	(202) 606-4248	(202) 606-4337
Tennessee Valley Authority (TVA)	(865) 632-3631	(865) 632-3412
U.S. Marshals Service	(202) 307-9100 (800) 336-0102	(202) 307-9177
U.S. Postal Service (USPS)	(202) 268-2000	(202) 268-2304 (202) 268-5211
U.S. Army Corps of Engineers (USACE)	(202) 761-1001 Via Army Opns Center (703) 697-0218	(202) 761-0378 (703) 693-6290

Non-Government Organizations

Organization	Telephone
Adventist Community Services	(770) 277-2525
American Public Works Association (APWA)	(202) 393-2792
American Radio Relay League, Inc.	(860) 594-0261
American Red Cross (Operations Center)	(703) 206-8822
American Society of Civil Engineers (ASCE)	(202) 789-2200
America's Second Harvest	(312) 263-2303 Ext 147
Association of General Contractors (AGC)	(202) 393-2040
Catholic Charities USA	(703) 549-1390 Ext 118
Christian Reformed World Relief Committee	(616) 224-0738
Church of the Brethren	(410) 635-8731
Church World Services	(212) 870-3154
Corporation For National Service (Disaster Services)	(202) 606-5000 Ext 124
Episcopal Church	(212) 716-6025
Friends Disaster Services	(330) 650-4975
Humane Society of The United States	(301) 258-3103/3119
International Association of Emergency Managers	(703) 538-1795
International Association of Fire Chiefs (IAFC)	(703) 273-0911
International Relief Friendship Foundation	(914) 366-0558
Lutheran Disaster Response	(773) 380-2822
Mennonite Disaster Service	(717) 859-2210
National Association for Search and Rescue (NASAR)	(703) 222-6277
National Emergency Management Association (NEMA)	(606) 244-8233
National Emergency Response Team	(888) 637-8872
National Organization for Victim Assistance	(202) 232-6682
National Voluntary Organizations Active in Disaster (NVOAD) (Executive Secretary)	(301) 890-2119
Nazarene Disaster Response	(407) 294-2917
Northwest Medical Teams, International	(503) 624-1000
Phoenix Society for Burn Survivors	(800) 888-2876
Points of Light Foundation	(202) 729-8161
Presbyterian Disaster Assistance	(502) 569-5797

Organization	Telephone
REACT International	(610) 434-3235
Salvation Army	(847) 795-3081
Southern Baptist Convention	(770) 410-6133
St. Vincent de Paul	(314) 576-3993
United Jewish Communities	(212) 284-6671
United Methodist Committee on Relief	(202) 548-4002
United States Service Command	(847) 689-2545
Volunteers in Technical Assistance (VITA)	(703) 276-1800 Ext 22
World Vision	(253) 815-2197

State Contact Directory

A

Alabama Emergency Management Agency
5898 County Road 41
P.O. Drawer 2160
Clanton, Alabama 35046-2160
(205) 280-2200
(205) 280-2495 FAX
ema.alabama.gov/

Alaska Division of Homeland Security and Emergency Management
P.O. Box 5750
Fort Richardson, Alaska 99505-5750
(907) 428-7000
(907) 428-7009 FAX
www.ak-prepared.com

American Samoa Territorial Emergency Management Coordination
(TEMCO)
American Samoa Government
P.O. Box 1086
Pago Pago, American Samoa 96799
(011)(684) 699-6415
(011)(684) 699-6414 FAX

Arizona Division of Emergency Management
5636 E. McDowell Rd
Phoenix, Arizona 85008
(602) 244-0504 or 1-800-411-2336
www.azdema.gov

Arkansas Department of Emergency Management
Bldg. # 9501
Camp Joseph T. Robinson
North Little Rock, Arkansas 72199-9600
(501) 683-6700
(501) 683-7890 FAX
www.adem.arkansas.gov/

C

California Governor's Office of Emergency Services
3650 Schriever Ave.
Mather , CA 95655-4203
(916) 845-8510
(916) 845-8511 FAX
www.oes.ca.gov/

Colorado Office of Emergency Management
Division of Local Government
Department of Local Affairs
9195 East Mineral Avenue
Suite 200
Centennial , Colorado 80112
(720) 852-6600
(720) 852-6750 Fax
www.dola.state.co.us/oem/oemindex.htm

Connecticut Office of Emergency Management
Department of Emergency Management and Homeland Security
360 Broad Street
Hartford, Connecticut 06105
(860) 566-3180
(860) 247-0664 FAX
www.ct.gov/demhs/site/default.asp

D

Delaware Emergency Management Agency
165 Brick Store Landing Road
Smyrna, Delaware 19977
(302) 659-3362
(302) 659-6855 FAX
www.state.de.us/dema/index.htm

District of Columbia Emergency Management Agency
2720 Martin Luther King, Jr. Avenue, S.E.
Second Floor
Washington, D.C. 20032
(202) 727-6161
(202) 673-2290 FAX
dcema.dc.gov

F

Florida Division of Emergency Management
2555 Shumard Oak Blvd.
Tallahassee, Florida 32399-2100
(850) 413-9969
(850) 488-1016 FAX
floridadisaster.org

G

Georgia Emergency Management Agency
P.O. Box 18055
Atlanta, Georgia 30316-0055
(404) 635-7000
(404) 635-7205 FAX
www.State.Ga.US/GEMA/

Office of Civil Defense
Government of Guam
P.O. Box 2877
Hagatna, Guam 96932
(011)(671) 475-9600
(011)(671) 477-3727 FAX
http://ns.gov.gu/

Guam Homeland Security/Office of Civil Defense
221B Chalan Palasyo
Agana Heights, Guam 96910
Tel:(671)475-9600
Fax:(671)477-3727
www.guamhs.org

H

Hawaii State Civil Defense
3949 Diamond Head Road
Honolulu, Hawaii 96816-4495
(808) 733-4300
(808) 733-4287 FAX
www.scd.hawaii.gov

I

Idaho Bureau of Homeland Security
4040 Guard Street, Bldg. 600
Boise, Idaho 83705-5004
(208) 422-3040
(208) 422-3044 FAX
www.bhs.idaho.gov/

Illinois Emergency Management Agency
2200 S. Dirksen Pkwy.
Springfield, Illinois 62703
Office: (217) 782-2700
Fax: (217) 524-7967
www.state.il.us/iema

Indiana Department of Homeland Security
Indiana Government Center South
302 West Washington Street, Room E208
Indianapolis, Indiana 46204-2767
Office: (317) 232-3986
Fax: (317) 232-3895
www.ai.org/sema/index.html

Indiana State Emergency Management Agency
302 West Washington Street
Room E-208 A
Indianapolis, Indiana 46204-2767
(317) 232-3986
(317) 232-3895 FAX
www.ai.org/sema/index.html

Iowa Homeland Security & Emergency Management Division
Department of Public Defense
Hoover Office Building
Des Moines, Iowa 50319
(515) 281-3231
(515) 281-7539 FAX
lowahomelandsecurity.org.

K

Kansas Division of Emergency Management
2800 S.W. Topeka Boulevard
Topeka, Kansas 66611-1287
(785) 274-1401
(785) 274-1426 FAX
www.ink.org/public/kdem/

Kentucky Emergency Management
EOC Building
100 Minuteman Parkway Bldg. 100
Frankfort, Kentucky 40601-6168
(502) 607-1682
(502) 607-1614 FAX
kyem.ky.gov/

L

Louisiana Office of Emergency Preparedness
7667 Independence Blvd.
Baton Rouge, Louisiana 70806
(225) 925-7500
(225) 925-7501 FAX
www.ohsep.louisiana.gov

M

Maine Emergency Management Agency
45 Commerce Drive, Suite #2
#72 State House Station
Augusta, Maine 04333-0072
207-624-4400
207-287-3180 (FAX)
www.state.me.us/mema/memahome.htm

Homeland Security and Emergency Management Division
Michigan Dept. of State Police
Michigan State Police
4000 Collins Road
Lansing, MI 48910
Office: (517) 333-5042
Fax: (517) 333-4987
www.michigan.gov/emd

CNMI Emergency Management Office
Office of the Governor
Commonwealth of the Northern Mariana Islands
P.O. Box 10007
Saipan, Mariana Islands 96950
(670) 322-9529
(670) 322-7743 FAX
www.cnmiemo.gov.mp

National Disaster Management Office
Office of the Chief Secretary
P.O. Box 15
Majuro, Republic of the Marshall Islands 96960-0015
(011)(692) 625-5181
(011)(692) 625-6896 FAX

Maryland Emergency Management Agency
Camp Fretterd Military Reservation
5401 Rue Saint Lo Drive
Reistertown, Maryland 21136
(410) 517-3600
(877) 636-2872 Toll-Free
(410) 517-3610 FAX
www.mema.state.md.us/

Massachusetts Emergency Management Agency
400 Worcester Road
Framingham, Massachusetts 01702-5399
(508) 820-2000
(508) 820-2030 FAX
www.state.ma.us/mema

Michigan Division of Emergency Management
4000 Collins Road
P.O. Box 30636
Lansing, Michigan 48909-8136
(517) 333-5042
(517) 333-4987 FAX
www.michigan.gov/msp/1,1607,7-123-1593_3507---,00.html

National Disaster Control Officer
Federated States of Micronesia
P.O. Box PS-53
Kolonia, Pohnpei - Micronesia 96941
(011)(691) 320-8815
(001)(691) 320-2785 FAX

Minnesota Homeland Security and Emergency Management Division
Minnesota Dept. of Public Safety
444 Cedar Street, Suite 223
St. Paul, MN 55101-6223
Office: (651) 296-0466
Fax: (651) 296-0459
www.hsem.state.mn.us

Mississippi Emergency Management Agency
P.O. Box 5644
Pearl, MS 39288-5644
(601) 933-6362
(800) 442-6362 Toll Free
(601) 933-6800 FAX
www.msema.org
www.msema.org/mitigate/mssaferoominit.htm

Missouri Emergency Management Agency
P.O. Box 116
2302 Militia Drive
Jefferson City, Missouri 65102
(573) 526-9100
(573) 634-7966 FAX
sema.dps.mo.gov

Montana Division of Disaster & Emergency Services
1900 Williams Street
Helena, Montana 59604-4789
(406) 841-3911
(406) 444-3965 FAX
dma.mt.gov/des/

N

Nebraska Emergency Management Agency
1300 Military Road
Lincoln, Nebraska 68508-1090
(402) 471-7410
(402) 471-7433 FAX
www.nema.ne.gov

Nevada Division of Emergency Management
2525 South Carson Street
Carson City, Nevada 89711
(775) 687-4240
(775) 687-6788 FAX
dem.state.nv.us/

Governor's Office of Emergency Management
State Office Park South
107 Pleasant Street
Concord, New Hampshire 03301
(603) 271-2231
(603) 225-7341 FAX
www.nhoem.state.nh.us/

New Jersey Office of Emergency Management
Emergency Management Bureau
P.O. Box 7068
West Trenton, New Jersey 08628-0068
(609) 538-6050 Monday-Friday
(609) 882-2000 ext 6311 (24/7)
(609) 538-0345 FAX
www.state.nj.us/njoem/index.html

New Mexico Department of Homeland Security and Emergency Management (DHSEM)
P.O. Box 27111
13 Bataan Boulevard
Santa Fe, New Mexico 87502
(505) 476-9600
(505) 476-9635 Emergency
(505) 476-9695 FAX
http://www.nmdhsem.org/

New York State Emergency Management Office
1220 Washington Avenue
Building 22, Suite 101
Albany, New York 12226-2251
(518) 292-2275
(518) 457-9995 FAX
www.semo.state.ny.us/

North Carolina Division of Emergency Management
4713 Mail Service Center
Raleigh, NC 27699-4713
(919) 733-3867
(919) 733-5406 FAX
www.dem.dcc.state.nc.us/

North Dakota Department of Emergency Services
P.O. Box 5511
Bismarck, North Dakota 58506-5511
(701) 328-8100
(701) 328-8181 FAX
www.nd.gov/des

O

Ohio Emergency Management Agency
2855 West Dublin-Granville Road
Columbus, Ohio 43235-2206
Office: (614) 889-7150
Fax: (614) 889-7183
ema.ohio.gov/ema.asp

Office of Civil Emergency Management
Will Rogers Sequoia Tunnel 2401 N. Lincoln
Oklahoma City, Oklahoma 73152
(405) 521-2481
(405) 521-4053 FAX
www.odcem.state.ok.us/

Oregon Emergency Management
Department of State Police
PO Box 14370
Salem, Oregon 97309-5062
(503) 378-2911
(503) 373-7833 FAX
egov.oregon.gov/OOHS/OEM

P

Palau NEMO Coordinator
Office of the President
P.O. Box 100
Koror, Republic of Palau 96940
(011)(680) 488-2422
(011)(680) 488-3312

Pennsylvania Emergency Management Agency
2605 Interstate Drive
Harrisburg PA 17110-9463
(717) 651-2001
(717) 651-2040 FAX
www.pema.state.pa.us/

Puerto Rico Emergency Management Agency
P.O. Box 966597
San Juan, Puerto Rico 00906-6597
(787) 724-0124
(787) 725-4244 FAX
www.gobierno.pr/AEMEAD/Inicio

R

Rhode Island Emergency Management Agency
645 New London Ave
Cranston, Rhode Island 02920-3003
(401) 946-9996
(401) 944-1891 FAX
www.riema.ri.gov

S

Tennessee Emergency Management Agency
3041 Sidco Drive
Nashville, Tennessee 37204-1502
(615) 741-4332
(615) 242-9635 FAX
www.tnema.org

Texas Division of Emergency Management
5805 N. Lamar
Austin, Texas 78752
(512) 424-2138
(512) 424-2444 or 7160 FAX
www.txdps.state.tx.us/dem/

T

Tennessee Emergency Management Agency
3041 Sidco Drive
Nashville, Tennessee 37204-1502
(615) 741-4332
(615) 242-9635 FAX
http://www.tnema.org

Texas Division of Emergency Management
5805 N. Lamar
Austin, Texas 78752
(512) 424-2138
(512) 424-2444 or 7160 FAX
http://www.txdps.state.tx.us/dem/

U

Utah Division of Emergency Services and Homeland Security
1110 State Office Building
P.O. Box 141710
Salt Lake City, Utah 84114-1710
(801) 538-3400
(801) 538-3770 FAX
www.des.utah.gov

V

Vermont Emergency Management Agency
Vermont Emergency Management Agency
Department of Public Safety
Waterbury State Complex
103 South Main Street
Waterbury, Vermont 05671-2101
(802) 244-8721
(802) 244-8655 FAX
www.dps.state.vt.us/

Virgin Islands Territorial Emergency Management - VITEMA
2-C Contant, A-Q Building
Virgin Islands 00820
(340) 774-2244
(340) 774-1491

Virginia Department of Emergency Management
10501 Trade Court
Richmond, VA 23236-3713
(804) 897-6502
(804) 897-6506
www.vdem.state.va.us

W

State of Washington Emergency Management Division
Building 20, M/S: TA-20
Camp Murray, Washington 98430-5122
(253) 512-7000
(253) 512-7200 FAX
www.emd.wa.gov/

West Virginia Office of Emergency Services
Building 1, Room EB-80 1900 Kanawha Boulevard, East
Charleston, West Virginia 25305-0360
(304) 558-5380
(304) 344-4538 FAX
www.wvdhsem.gov

Wisconsin Emergency Management
2400 Wright Street
P.O. Box 7865
Madison, Wisconsin 53707-7865
Phone: (608) 242-3232
Fax: (608) 242-3247
emergencymanagement.wi.gov/

Wyoming Office of Homeland Security
122 W. 25th Street
Cheyenne, Wyoming 82002
(307) 777-4900
(307) 635-6017 FAX
wyohomelandsecurity.state.wy.us

Updated: 23-Jul-2008 (307) 635-6017 FAX
http://wema.state.wy.us

Disaster Recovery Policies, Procedures and Forms Index - 600

Disaster Recovery Policies, Procedures and Forms Index

Disaster Recovery Policies, Procedures and Forms Index

[This page intentionally left blank]

www.ingramcontent.com/pod-product-compliance
Lightning Source LLC
LaVergne TN
LVHW062243200926
848484LV00006B/1599

* 9 7 8 1 9 3 1 5 9 1 4 0 9 *